Climate Induced Migration:
Assessing the Evidence from Bangladesh

Dr. Sohela Mustari

Assistant Professor
Department of Sociology and Anthropology
Kulliyyah of Islamic Revealed Knowledge and Human Sciences
International Islamic University Malaysia

PARTRIDGE

Copyright © 2022 by Dr. Sohela Mustari.

ISBN:	Hardcover	978-1-5437-7083-4
	Softcover	978-1-5437-7081-0
	eBook	978-1-5437-7082-7

All rights reserved. No part of this book may be used or reproduced by any means, graphic, electronic, or mechanical, including photocopying, recording, taping or by any information storage retrieval system without the written permission of the author except in the case of brief quotations embodied in critical articles and reviews.

Because of the dynamic nature of the Internet, any web addresses or links contained in this book may have changed since publication and may no longer be valid. The views expressed in this work are solely those of the author and do not necessarily reflect the views of the publisher, and the publisher hereby disclaims any responsibility for them.

Print information available on the last page.

To order additional copies of this book, contact
Toll Free +65 3165 7531 (Singapore)
Toll Free +60 3 3099 4412 (Malaysia)
orders.singapore@partridgepublishing.com

www.partridgepublishing.com/singapore

CONTENTS

Acknowledgement xi

Chapter 1   Introduction and Overview 1

                Climate change and extreme weather events 2
                Climate change and poverty 3
                Climate Change and Migration 4
                Climate change and migration:
                Old context in new concern 5
                Global scenario of climate change and migration 6
                Research Questions 7
                Research Objectives 7

Chapter 2   Literature Review 8

                Introduction 8
                Climate Change and society 9
                Migration: Theoretical Framework 11
                Why do people migrate? 13
                Migration in the context of vulnerability and
                adaptation to climatic variability and change 14
                Other coping and adaptation strategies 14
                Conceptual Framework 15

Chapter 3   Focus Countries and Data 16

                Focus country 16
                Criteria for the choice of focus country 17

Chapter 4	Data Collection 19
	1. Survey 19
	2. Case Study 20
	3. Interview 21
	4. Focus Group Discussion 21
	Conclusion 21
Chapter 5	Results and Findings: Perception of The Villagers 22
	Introduction 22
	Perceptions on climate change 23
	Impact on households and
	Socio-economic disturbances 25
	Conclusion 29
Chapter 6	Results and Findings: Causes of Migration 30
	The causes and factors of migration 30
	Conclusion 34
Chapter 7	Results and Findings: Migration and Migrants 35
	Migration and migrants related information 35
Chapter 8	Results and Findings: Migrants Experience 44
	Introduction 44
	Conclusion 65
Chapter 9	Results and Findings: Experiences of Household of Origin 66
	Introduction 66
	Conclusion 73

Chapter 10 Discussion, Conclusion and Recommendations 74

    1. Socio-demography and migration: 74
    2. Migration throughout their life/ migration generation after generation: 75
    3. Chiefly internal movement seeking for a better standard of living 75
    4. Typically with the expectation of eventually returning, again highlighting circular migration 76
    5. Voluntary or Distressed and forced migration: 77
    6. Environmental reasons more generally could possibly trigger a decision to move 77
    7. Migration and social network 78
    8. The social valuation of migration and migrants in sending society is higher than in receiving society 78
    9. Crisis of social identity in the place of receiving areas: psychological separation and competition in new areas 79
    Conclusions 79
    Limitations 80
    Recommendations 81

Reference 83

Illustrations

UNFCC: United Nations Framework Convention on Climate Change
GHG: Green House Gas
IPCC: International Panel on Climate Change
IOM: International Organization for Migration

Figures

5.1: **Type of natural disasters in the last 3 years.**
Data source: Field survey, 2017..25
7.1: **Type of Migration,** Data source: Field survey, 2017 38
7.2: **Monetary help from migrated family members,**
Data source: Field survey, 2017..39
7.3: **Types of work done by the migrated family
members,** Data source: Field survey, 2017 39
7.4: **Reasons for their migration,**
Data source: Field survey, 2017..41
7.5: **Type of migration conducted by family
members,** Data source: Field survey, 2017 42
7.6: **Reasons of not migrating with full family,**
Data source: Field survey, 2017..43

Tables

5.1: Sign of climate change in own area 23
5.2: Type of natural disasters experienced
in the last 3 years..24
5.3: Types of vulnerabilities due to natural
disasters in the last 3 years...25
5.4: Amount of food production 27
5.5: The reasons of less food production than before.........28
6.1: Techniques of adjustment with the
entire socio-economic crisis 31
6.2: Techniques of adjusting to a food crisis......................32

6.3: Techniques to adjust with the fuel crisis.......................33
7.1: Number of migrated family members 37
7.2: Amount of money (per month)..40
7.3: Relationship with the migrated member......................41

Diagram

**2.1: Conceptual Framework of Climate Change
outputs and migration** 15
7.1: Black, et al; (2011). Slightly modified. 37

ACKNOWLEDGEMENT

This book owes a lot to the people of Bangladesh who took part in this research, giving time and energies to ensure that their voice will be heard across the world. The respondents of all ages and gender participated in this research willingly and gave the information so that the whole world can hear their voice.

This research was made possible by a research fund of Institute of Research and Training (IRT) of Southeast University, Bangladesh and by a donation of an anonymous businessman who has great enthusiasm for research and education. I especially want to thank my research team for their invaluable assistance in site visiting with me and collecting data with survey from these remote areas.

My thanks go to the government and non-government organizations and their representatives both in Dhaka and the research areas for giving me information, security and valuable feedback on my research.

To my ex-colleagues of Southeast University, Dr. Mehe Zebunnesa Rahman, Dr. Farhana Ferdausi and Dr. Tanvir Mahmud for their vital support during the extraordinary three-year (end of 2017 to 2020) of this research by guiding, reviewing and sharing their knowledge and wisdom. Finally, to my husband, parents and kids for their support and sacrifices- my sincere thanks to all of them.

CHAPTER 1

Introduction and Overview

Over the last few years, researchers and policy makers are showing their interest in measuring the impact of climate change on migration and human displacement. It is projected that by the next 40 years, around one billion of the world population may move from their own place that has been affected by climate change (Laczko & Aghazarm, 2009). Already it is evident that some parts of the world have become less inhabitable than other parts due to climate change and its effects, such as desertification, pollution, natural disasters or decline of agricultural productions.

There is a tendency for researchers and policy makers to see migration from a negative perspective and to initiate policies to reduce migration numbers from disaster prone areas. However, it would be oversimplification if it is said there are no such works. Few studies are found where researchers claim that migration could be one of the alternatives when other alternatives do not function properly to reduce vulnerability from climate change (Mustari & Karim, 2017). Despite the interest of researchers and policy makers to understand migration from climate change perspectives, very few documents are available in Bangladesh that answer the following questions:

How many are migrating due to climate change? Who are migrating? When and where are they migrating? Are the new destinations better than their origin? Is the migration temporary or permanent? Internal or international? What are the consequences of migration for the people who move, for those left behind? These questions are tried to be answered properly with much concern so that policy makers can initiate the required policies to reduce the vulnerability of the community.

Climate change and extreme weather events

According to UNFCCC, "Climate change" means a change of climate which is attributed directly or indirectly to human activity that alters the composition of the global atmosphere and which is in addition to natural climate variability observed over comparable time periods. The Convention on Climate Change sets a general framework for international exertions to face the challenges set by climate change. It recognizes that the climate system is a common resource which constancy can be exaggerated by industrial and other discharges of carbon dioxide and other greenhouse gases. However, climate change has heterogeneous impacts based on geographic differences. Thus, some areas are more vulnerable than others and obviously some countries are more responsible in generating carbon gases than others. In response to UNFCCC, the Kyoto Protocol (1997) emerged, which is an international agreement linked to the UNFC on climate change. In this agreement all parties agreed to set a target in binding Green House Gas (GHG) emission. In the Kyoto Protocol, it is specifically mentioned that the developed countries are principally responsible in generating GHG but the sufferers are mostly the less developed countries. So, this Protocol decided to monitor the countries strongly so that they can control the amount of GHG.

The Paris Agreement (2015) aimed to strengthen the global response to reduce the threat of global climate change. To reduce the vulnerabilities of climate change and to take proper initiatives of adaptation, all nations came to a common platform to assist the developing countries. To make the Paris Agreement successful, true guidelines, modalities and other necessities were taken in 2016. Some policies like to strengthen the knowledge, technologies, practices and efforts of local communities and indigenous peoples which are initiated and focused during this period. However, the UN Climate Change News (3 July 2018) stated that climate change vulnerabilities are increasing with the cost of capital and debt payments. This report also mentioned that the situation is worse in less developed countries than others (United Nations Climate Change, 2018).

Climate change and poverty

Climate change is a big threat for poor people of any country to achieve the goal of sustainable development. The poor people in poor and less developed countries are more vulnerable to any kind of natural hazards resulting from climate change. These people are threatened by natural disasters by losing assets and livelihoods, suffer from diseases, and loss of infrastructure. Climate change and its effects with huge droughts, floods, reduced rainfall, food scarcity from less production, and high rise of food prices cause vulnerability not only for the poor but also other members of the society. This is how climate change creates a new poor every year and they become a burden for the society as well as country (Hallegatte et al; 2015).

The long-term effect of climate change on the poor is acute than the short term effects. If the proper policy is not taken to reduce global warming or the effects of global warming and natural disasters, poor people will lose their livelihoods and soil will lose their productivity. On the other hand, if the policy is taken to stop global warming and to reduce emissions, huge money needs to be invested which eventually will increase the price of food items. So, in both ways, climate change threatens the poor of developed countries and the people of poor and less developed countries (Baker, 2012).

Millions of the global population depend on agriculture for their livelihood and subsistence which is critically under threat with the climatic variability (Skoufias, 2012). The fact is, the poor are poor because they have less resources and capacity, and less support from the relatives, neighbours and community. Likewise, they have less social safety to cope, adapt and be resilient with climate change. So, this group of people remain vulnerable with climate change and its effects and it becomes harder to achieve sustainable development.

Climate change forms long term menaces for the society. Apparently, it seems that short term strategies such as financial aid or food aid can bring the victims out of poverty but in long term analysis it is found these strategies are not suitable. As a result, in most cases, households from the community fall into poverty. In addition, hazardous weather pushes the people to withdraw from any kind of investment because of the high risk of loss from disaster (Kron, 2013). Moreover, these people choose low investment with low profit activities only to keep themselves out of risk. This behavioral pattern keeps the community in poverty.

Climate Change and Migration

Climate change is extensively accepted to be one of the greatest threats to the contemporary global community (Cameron, 2018). Global impacts of climate change will increase in the near future by increasing the number of mortality and water borne diseases, decreasing sources of livelihoods and marginalized the poor (IPCC, 2014). Globally, climate change has unequal effects on different countries; similarly, climate change has unequal effects locally.

Migration is one of the adaptation alternatives that people choose to cope with climate change (Dell, 2014).

Climate change and migration: Old context in new concern

Migration to other suitable places as a result of climate change is not a new issue of discussion. For centuries, people move from one place to another to adjust with the changed environment to survive and look for foods and livings. Throughout the progression of civilization, socio-archeologists claimed that human settlement patterns were determined by changed climate. Scarce resources and dense populations were the key push factors of migration and pull factors in developing the first civilization beside the riverine areas. Evidence also suggests that population movement from a colder region to a milder region by the Visigoths was due to their survival from the extreme climate. Likewise, people of pastoral societies moved with their animals in search of water and food for their belongings. So, the link between climate change and migration is not a new issue of concern, rather, it has a past context from ancient times.

Historically, migration is a strong mechanism of coping with climate change. Though evidence showed people of pastoral society moved to other places looking for food and water and took place beside the river, but this movement is not limited only for the nomadic people. Even after the permanent settlement, people moved to other places if the climate changed and its effects become extreme to survive. Researchers and policy makers show their concern for migration as the permanent settlement of humanity are at stake with climate change. As data shows that climate change concurs stress in socio-economic life which pushes for forced migration from rural to urban areas. In most cases, the migration is to earn a living and as a temporary basis to cope with the climate change and its effects. As they do not have enough capital due to poverty, they cannot go for long distance migration; rather, they conduct a short distance migration to a nearer city. The International Organization of Migration (IOM, 2008) reports, in most of the West African countries, migration takes place on a temporary basis and to a short distance.

Global scenario of climate change and migration

Sea level rise, deforestation, land degradation or/and natural disasters pose a threat to the global society by causing negative effects on their livelihoods, agricultural production, health and infrastructural development. Researchers predicted that the socio-economic burden caused by environmental stress will lead to a large scale of population moving to a better place in the near future (Myers & Papageorgiou, 1997; Stern et al., 2006). Research has shown that a single event of natural disaster such as flood or drought may cause a limited number of migrations, but the long-term effects of climate change push a larger number of the population to migrate to a safer place. Climate change with less agricultural production, water scarcity, and unemployment causes the population to think of a long-distance migration instead of short distance migration (Henry et al., 2004). However, in all cases, migration is not the first choice of households to cope with climate change. In both situations (either the single event of a disaster or the long-term effects of climate change), households prefer to use the existing means of survival such as selling lands or livestock, looking for loans from formal and informal sources, waiting for government or non-government support etc. If all the existing means fail, household members will migrate. Migration is the least preferable alternative to the population as it requires money, social network and other social support which is very difficult for many of the population.

For the abovementioned discussion, in this book the following objectives are achieved with the answers to the following questions.

Research Questions

1. How does climate change create insecurity in rural people's life?
2. How does climate change motivate rural people to commit to migration?
3. How does migration cause rural people to elevate their social insecurity?

Research Objectives

1. To know the nature and form of insecurity that rural people experience due to climate change and its impact.
2. To know the causes and factors that motivate rural people to take part in migration.
3. To know the role and contribution of urban migration in elevating the insecurity of the villagers.

CHAPTER 2

Literature Review

Introduction

Bangladesh is one of the most populated countries in the world which has caused the nation to face poverty extensively. Specifically, this poverty is the principal reason for vulnerability instead of being resilient against climate change and its effects (Adger, 2006; Schipper & Pelling, 2006). The geographical setting of Bangladesh forces the country to experience multifaceted natural hazards throughout the year including droughts, cyclones and floods. Uddin (2009) projected the impact of climate change on Bangladesh to be visible through floods, flashfloods, water logging, losing land, drainage congestion, heavy spells of downpour, salinity, surface water, shortage of water, moisture stress, droughts, lack of drinking water, and difficulties in harnessing domestic water. Thus, the implication of all these natural hazards will impact Bangladeshi agriculture, water infrastructure, rural-urban infrastructure, human health, and vulnerable ecosystem and livelihood groups.

Secondary sources (e.g., Koenig, Ahmed, Hossain &Mozumder, 2003; Saroar & Routray, 2010) claimed that the Bangladeshi rural community is conservative in character and thus, rural inhabitants in general avoid moving elsewhere, preferring to adjust to the changed situation in the climate and the effects it has on them. However, if there is no other alternative and if the inhabitants need to move, they prefer to do temporary internal migration rather than international migration which is very rare for these people.

The poor and technologically backward people of Bangladesh have very limited alternatives to cope with these natural hazards. Among them, migration is one of the preferred alternatives for the climatically insecure people of Bangladesh. World Bank (2013) states in their report that climatic shocks are gradually threatening rural livelihoods which forcefully send people to migrate with new hope and aspiration to urban areas.

Climate Change and society

Climate change is an intangible issue which is explicable only through its impacts (Rehdanz, Welsch, Narita & Okubo, 2015). At this point, it is precisely understandable that climate change and its effects come gradually; it does not happen all of a sudden. To adjust to the changes in the climatic situation, previous literature shows that some members of a family migrate to other places where they can make money and send it back to their families. However, researchers and policy makers have only started to conduct their research on migration as the consequence of climate change in the last two decades (Perch-Nielsen, Bättig & Imboden, 2008).

A South African based study by Tibesigwa and Visser (2015) tested the food security level of different gender households in rural areas. They found that due to climate change, female-headed households suffered greater than male-headed households. This is because of the females' incapacity to migrate to other places to manage food. Another work by Barrios, Bertinelli and Strobl (2006) which was based on sub-Saharan Africa shows that although the rural economy of that country was based on agriculture, where the dependency has been declining gradually since the 1950s as rainfall has decreased a lot since then. This agriculturally dependent population is looking for urban migration for their new settlements. In their words, they asserted "climate change scenarios tend to suggest that extreme climate variations are likely to cause abrupt changes in human settlements and urbanization patterns in the sub-Saharan Africa more than anywhere else in the world". A household survey by Cattaneo and Massetti (2015) in Ghana and Nigeria tried to ascertain the role of migration in employing adaptation to climate change. They found that in both countries there was a significant relationship between migration and farm-based household. On the other hand, they did not find any significant relationship between migration and non-farming household. Reuveny (2007) also discussed rural-urban migration in China where climate change and its effect was so prominent with regular attacks of floods, land degradation, droughts and subsequent water scarcity that 20-30 million rural people were forced to migrate to urban centres from nearby provinces.

In the academic thesis written by Plowman (2015), climate change and its effects have been found to work as a push factor for Bangladeshi rural people who work mostly in the garment sector with a cheap wage to survive on. In a recent paper, Iqbal and Roy (2015) stated that climate change impact causes changes in temperature and rainfall which push the farmers to look for new working areas, usually in a nearby urban area. Rural-urban migration is now a well-accepted process of bringing success in life even in a less developed and remote society. Brueckner and Lall (2015) analysed a few factors of migration which work either as the pull or push factor of migration where climate change is deemed as one of the push factors. They add further that the trend of migration from rural to urban areas due to climate change issues is greater in developing countries compared to developed countries. Some researchers (e.g., Paul, Hossain and

Ray, 2013) stated that due to climate change effects in Bangladesh, vulnerable groups, especially from "monga" prone areas include the agricultural wage laborers, marginal farmers, female-headed households and other marginal groups of the society. Their research also mentioned a few coping strategies like borrowing money, selling of assets and migrating.

Finally, Hunter, Luna and Norton (2015) stated the relationship between environmental change and migration issues can be described in a multi-disciplinary way. However, they further mentioned that sociological research needs to move away from the push and pull factors related research only; that migration should be looked at from other sociological perspectives like social network, socio-economic condition and others.

Migration: Theoretical Framework

Early writers like Fairchild (1925) saw immigration differently from invasion, conquest and even from colonization. From over a century, scholars and researchers have been providing general explanations of human migration. From the academic disciplinary ground, most of the explanations are linked with either economics, sociology or geography. Migration was not the primary concern of their analysis, rather, these academic disciplines used migration to explain other human behaviors in the society (Arango, 2017).

Most of the early theories of migration focused on economic factors. However, in the age of modernization and globalization, migration is seen as the rebalancing of man-power in all societies (Todisco, Brandi, & Tattolo, 2003). There are a number of models of migration decision making. One of the migration models is an economic model which suggests that before migrating the potential migrants and evaluates the cost and benefits of his migration (Massey et al; 1993). If the cost and benefits are higher in local areas, the potential migrants will stop himself from moving. On the other hand, if the cost and benefits are higher in the destination place, h/

she will go for migration. Thus, according to the economic model, cost and benefits (financial and otherwise) work as the push and pull factors for migration. A similar model was given by Klaiber (2014); however, unlike many other models he gave a hypothesis by relating household migration and climate change. He stated that household migration occurs through: (1) changes in economic opportunities from climate change, and (2) climate facilities resulting from climate change (Klaiber, 2014).

Migration can be the a temporary or permanent choice of movement by an individual or groups of people from one geographical location to another. The reason for this movement may vary ranging from better livelihood to political instability (Hagen-Zanker, 2008). Ravenstein (1889), in his "The laws of migration" provided an empirical but incomplete explanation of modern migration. He expressed that the origin of migration is economical and mostly directed to short travel distances.

By using a rational choice approach and economic sociology of migration, this model (Haug, 2008) stressed on the role of social network. This model termed social network as the social capital of potential migrants which influence them to make a migration decision. In case of criticism of "migrant network", more explanations on migration were shown. This approach showed that beyond the network, migrants need other institutional support when migrating. Moreover, this approach stressed that this network is not worthy for all kinds of migrants equally. For example, the explanation of network theory is not equally true for all types of migrants, e.g., between students and labor. There are some models such as Tabor and Milfont (2011) which showed migration as the voluntary action of the actors. In this socio-psychological model, they proposed four stages to complete the voluntary migration. This model starts with intrapersonal factors that motivates potential migrants to conduct migration. This model ends with the acculturation stage where the individual migrants socio-psychologically adjust themselves with the changed environment.

Migration is defined as the permanent or semi-permanent alteration of dwelling. Parkins' (2010) theory of migration talked about the push and pull factors of migration. According to his idea, people change their residence either by prompting them to move or by forcing them to move. With some other factors, he mentioned a lack of socio-economic opportunities is the push factor in forcing people to move. Consequently, a suitable place with more opportunities attracts people to move.

Wallenstein in his World System Theory stressed that migration is a means of exploitation of the rich to gain cheap labor for the uneven development (Coccia, 2018). However, unlike this labor migration, another migration is seen in his work. He showed the migration of skilled workforce as the brain circulation between two communities. This circulation brings profit for both parties - the skilled migrants and the receiving communities.

Why do people migrate?

People migrate for many reasons. From the historical period, economics and related literature relate migration with higher income opportunities. Before migrating, an individual calculates his socio-economic benefits and selects his destination from the alternative choice. Migration works like an investment for them (Sjaastad, 1962). This investment has a relationship with demographic factors, for instance, age. People prefer to send young people in their early age for migration and are usually either the head of the family or the eldest son of the family (Castro & Rogers, 1984). Besides economic explanations of migration, there are non-economic explanations too (Uhlenberg, 1973). A sociological explanation for transnational migration was given by Castles (2003). Internal displacement or the development induced displacement forces the individual to look for migration. Similarly, if an individual has a strong social network, he/she may choose to migrate (Castles, 2003). In changing the environmental situation, migration can be considered as one of the alternatives for adaptation by individuals or the households (McLeman & Hunter, 2010).

Migration in the context of vulnerability and adaptation to climatic variability and change

The unpredictable changing nature of climate and climatic risk put human adaptability in an unprecedented challenge. In recent years, although people have become ever more conscious of the threats that climate change enacts to humanity, vulnerability resulting from climatic stress is dynamic in nature, depending on relations with power, resource distribution, knowledge, and technological development (Eakin, 2005). Local experience of social vulnerability differs in nature and their opportunities for adaptation likewise (Sen, 1981). The farming and livestock-based society of West African Sahel is vulnerable to climate change and its effects as environmental degradation changes the rate of rainfall (Van der Land & Hummel, 2013).

In changing the environmental situation, migration can be considered as one of the alternatives for adaptation by individuals or households (McLeman & Hunter, 2010). Like vulnerability, different people from different socio-demographic background prefer to migrate differently. Migration decision is influenced not only by environmental circumstances but also by cultural, economic, political, and social circumstances (Van der Land & Hummel, 2013). All these can work as the pull or push factors for the migration of an individual.

Other coping and adaptation strategies

An overpopulated society with weak infrastructure and poor livelihood make the inhabitants to be highly vulnerable in natural disasters (Braun & Aßheuer, 2011). A fast population growth impulses more and more people to take shelter in unprotected places such as near the rivers and water bodies to be resilient from natural hazards (Burkart, et al; 2008).

Climate change issues get high attention from the global audience due to the low capacity of societal people to cope with natural disasters like low rainfall, floods, and extreme climatic events. Adaptation capacity varies from individual to individual, household to household. Coping strategies for climate change operate at different levels, such as individual / household level, community / neighborhood level, and institutional / citywide / village wide levels (Jabeen, H., Johnson, C., & Allen, A. (2010). Empirical research found that relatives and networks play vital roles for individuals and groups to cope with the disasters, like floods (Braun & Aßheuer, 2011). Another empirical research with the farmers showed that food for work projects, depending on credit, engaging in petty businesses, and decreasing the amount and frequency of daily meals are some of the alternatives that villagers from Ethiopia took to adapt to climate change and its effects (Mengistu, 2011).

Conceptual Framework

Conceptually and methodologically, this research began its discussion depending on the sociological theory. Thus, as a unit of analysis, it considers the individual, household, event, migrant discourse and the village and city itself. It is considered in this research that internal migration has increased in rural Bangladesh as a consequence of climate change.

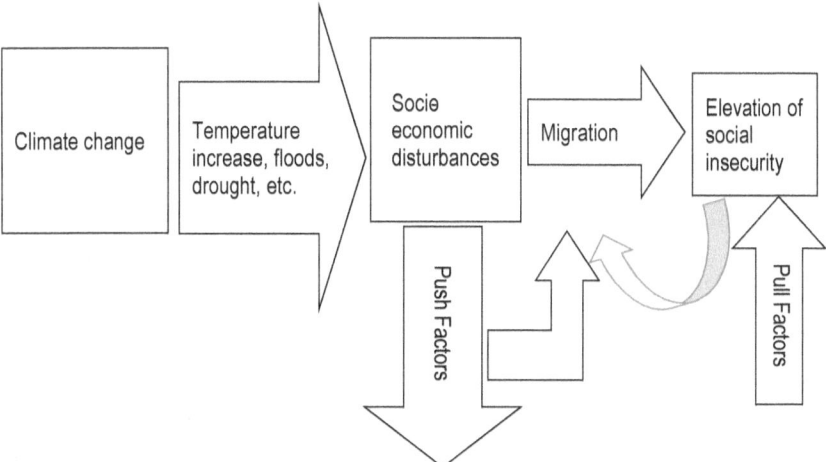

Diagram 2.1: Conceptual Framework of Climate Change outputs and migration

CHAPTER 3

Focus Countries and Data

Focus country

Bangladesh is one of the most populated countries in the world which has caused the nation to face poverty extensively. Specifically, this poverty is the principal reason for vulnerability instead of being resilient against climate change and its effects (Adger, 2006; Schipper & Pelling, 2006). The geographical setting of Bangladesh forces the country to experience multifaceted natural hazards throughout the year including droughts, cyclones, and floods. Uddin (2009) projected the impact of climate change on Bangladesh to be visible through floods, flashfloods, water logging, losing land, drainage congestion, heavy spells of downpour, salinity, surface water, shortage of water, moisture stress, droughts, lack of drinking water, and difficulties in harnessing domestic water. Thus, the implications of all these natural hazards will impact Bangladeshi agriculture, water infrastructure, rural-urban infrastructure, human health, vulnerable ecosystem and vulnerable livelihood groups.

Secondary sources claimed that the Bangladeshi rural community is conservative in character and thus, rural inhabitants in general avoid moving elsewhere, preferring to adjust to the changed situation in the climate and the effects it has on them (Koenig, Ahmed, Hossain & Mozumder, 2003; Saroar & Routray, 2010). However, if there are no other alternatives and if the inhabitants need to move, they prefer to do temporary internal migration rather than international migration which is very rare for them.

The poor and technologically backward people of Bangladesh have very limited alternatives to cope with these natural hazards. Among them, migration is one of the preferred alternatives for the climatically insecure people of Bangladesh. World Bank (2013) stated in their report that climatic shocks are gradually threatening rural livelihoods which forcefully send people to migrate to urban areas with new hope and aspiration.

Criteria for the choice of focus country

Bangladesh context

Bangladesh is widely declared as the most vulnerable country of climate change. Climate change and its effects through the changes in temperature and rainfall, amplified amount and severity of floods and droughts, cyclones, storm surges and salinity distress the communities, ecology and infrastructure of the country (Huq & Rabbani, 2011).

Although in recent times, Bangladesh has progressed its economy and social standards, but still a huge population is living below the poverty line. This population is vulnerable to climate change and unable to cope with its effects. Secondary sources claimed (e.g,. Alam et al., 2011) that more than 50 million people are affected by the climate change and its effects in every five years.

Nevertheless, awareness of the common people about climate change has increased to a great extent than even the developed countries, but the policies of climate change are still lagging behind (Ayers & Huq, 2009).

Context of Study area

Gaibanda was chosen because seasonal natural disasters strike the place almost every year. The disasters are called *"monga"* which is a new concept to explain the shocks due to climate change and its effects (Zug, 2006). Zug defines *monga* as "a seasonal food insecurity in ecologically vulnerable and economically weak parts of north-western Bangladesh, primarily caused by an employment and income deficit before *aman 2* is harvested. It mainly affects those rural poor, who have an undiversified income that is directly or indirectly based on agriculture" (2006: p, 2).

Another district, Sirajganj, was chosen because every year it is overflowed by river water that causes flood. Baki and his co-researchers (2015) described the Sirajganj district in this way; "Sirajganj district lies on the bank of the most treacherous river Jamuna. The monsoon spillage of Jamuna is so high that it regularly overflows the banks and creates flooding in most of the upazilas of Sirajganj district" (Baki et al., 2015: p, 103). The third district, Bogra, was chosen for the same reason as Sirajganj. Some parts of this district are very much vulnerable to floods and riverbank erosion. Therefore, these three districts were chosen purposively as they can justify how the environmentally vulnerable people living there suffer from climate change and choose migration for their survival.

CHAPTER

4

Data Collection

The research was conducted mostly in the rural areas of Bangladesh. For this research, three villages from the districts of Sirajganj, Bogra, and Gaibanda from the greater north of Bangladesh were chosen purposively. These districts face climate change effects through various natural disasters like floods, droughts, riverbank erosion and others throughout the year.

1. Survey

A short and simple survey was conducted with the household heads in all three villages. A total of 384 samples (the sample size of 384 is accepted for any population size as cited in Krejcie, & Morgan, 1970) were taken with 128 samples taken from each village. These 384 households were chosen through a simple random sampling process and a questionnaire was used to collect data from these household heads. To avoid the duplication of respondents, each house was given a code. The collected data was analysed through descriptive statistics.

2. Case Study

Apart from this survey, a total of 16 case studies were conducted with potential respondents. The case studies were with two different categories of respondents.

   a. A total of four (4) case studies were done with villagers who are suffering from climate change effects and have sent at least one family member to the urban areas to ease their sufferings. These respondents were chosen purposively so that information about their migrated family members and their contributions to their family life can be examined. A checklist was prepared for these interviews and a thematic procedure was utilized to analyze the collected data.
   b. Apart from this, a total of twelve (12) case studies were conducted with migrated villagers who are residing temporarily or permanently in urban areas due to climate change effects. These interviews examined both the pull and push factors of internal migration to urban areas. In addition, these interviews fulfilled the objective of getting information on how urban areas are helping the climatically insecure people in making the adaptation.

A checklist was prepared for these interviews and a thematic procedure was utilized to analyze the collected data.

3. Interview

Finally, the opinions from experts were gathered with five in-depth interviews. These five experts were chosen purposively. They were asked about the general situation of climate change and its effects on these northern villages of Bangladesh. They focused on the rural-urban migration situation and its role in overcoming the villagers' crisis.

A checklist was prepared for these interviews and thematic procedures was utilized to analyze the collected data.

4. Focus Group Discussion

To increase the research's validity and reliability, a total of three focus group discussions (FGD) were conducted in the three selected villages. One FGD was done in each village. Six to 12 villagers participated in these discussions. However, the participants were selected conveniently. People from different genders joined in these discussions and shared their experiences on natural disasters and migration.

Conclusion

To increase the reliability and validity of the research, the triangulation method was used from different perspectives. To escalate the reliability of this research, more than one research area was used. Moreover, to get authentic data, different data collection tools were also used. The used methods were verified with the secondary sources and legitimated that these used tools are authentic in this kind of research.

CHAPTER 5

Results and Findings: Perception of The Villagers

Introduction

Climate change is a difficult issue to identify easily based on personal experience. Moreover, climate change becomes a more complicated issue with the continuous debate of general people, scientists, academics and politicians on climate change variability (Weber, 2010). The debate on climate change is not only for strategic and political reasons but also for psychological and cultural reasons. Even though climate change has long term effects on socio-economic life, general people are less concerned about it. Nevertheless, it is very much required for them to be aware of it for their further preparation for future climatic risks.

Climate change is the statistical average of weather of a region. Public perception on climate change differs over time (Capstick et al; 2015). With laymen observation and faulty memory, their experience of climate change is not always acceptable. So many researchers (e.g., Weber, 2010, 2016) claimed that people's perception with the description on climate change is not always acceptable but rather numerical and graphical presentations are more scientific and worthy. Other scientific research (e.g., Mertz et al; 2009) used both qualitative interviews and household surveys to know the perception on climate change. In this chapter, villagers' experience of climate change is presented with numerical and graphical data.

Perceptions on climate change

To know the nature and form of insecurity that rural people experience due to climate change and its impacts.

Table 5.1: Sign of climate change in own area

Sign of climate change in own area	Frequency	Percentage (%)
Lack of rain	193	50.3%
Unusual amount of heavy rain	299	77.9%
The weather is hotter than before	289	75.3%
Changing river banks	325	84.6%
Lack of water for irrigation	129	33.6%
Land fertility gone down	161	41.9%
Arsenic in the water	113	29.4%
More molest from unwanted pests	105	27.3%
Frequently storms	211	54.9%

*multiple answers were counted. Data source: Field survey, 2017

The villagers are mostly illiterate, and they are not concerned of the term "climate change". However, informal discussions with the villagers revealed that a number of non-governmental organizations are working with these villagers to make them aware of climate change. To understand the severity of climate change in their daily life, in this research, they were asked to identify the signs of climate change that they see in their own areas. As the research areas were purposively chosen and all three villages reside beside the rivers, very naturally 84.8% of the respondents stated (Table 5.1) that due to the impact of climate change their river banks are changing continuously. Next to it, 77.9% of the respondents said that due to climate change and its effect, an unusual amount of heavy rain occur every year. According to them, this is one of the major causes of flood in their areas. The villagers are considering the weather as extreme as 75.3% of the respondents stated that the weather is hotter than during their childhood time. According to them, winter season comes late nowadays and it remains hotter than before. Although the villagers considered an unusual amount of heavy rain happens in those days, at the same time, around half of the respondents (50.3%) deliberated that the lack of rain in that season is also a sign of climate change in their areas. Due to lack of rain, instead of being fertile, the lands transformed into deserts. Remarkably, this survey found that arsenic contamination in water was considered as one of the least (29.4%) sign of climate change in these northern parts of Bangladesh.

Table 5.2: Type of natural disasters experienced in the last 3 years

Type of disasters	Frequency	Percentage (%)
Floods	369	96.4%
Riverbank erosion	350	91.1%
Land erosion	235	61.2%
Droughts	90	23.4%
Storms	187	48.7%

Data source: Field survey, 2017

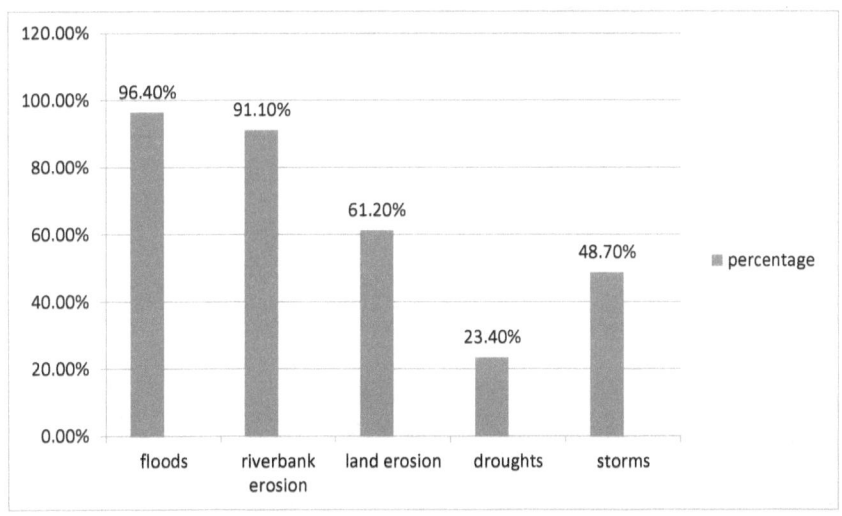

Figure 5.1: Type of natural disasters in the last 3 years. Data source: Field survey, 2017

In the Figure 5.1, we can see in the last three years that villagers of these areas mostly (96.40%) experienced floods. Next, riverbank erosion (91.10%) and land erosion (61.20%) were the highest experienced disasters by the villagers. Although in other parts of the survey the villagers proclaimed that the temperature was hotter and the rain was lesser than before, but in this section the least (23.40%) experienced disaster in the last three years was drought.

Impact on households and Socio-economic disturbances

Table 5.3: Types of vulnerabilities due to natural disasters in the last 3 years

Types of difficulties	Frequency	Percentage (%)
Degradation of land quality	209	54.6%
Drinking water scarcity	304	79.2%
Land pattern change	227	59.1%

Labor force pattern change	140	36.5%
Extinction of bio-diversity	81	21.4%
Food habit change	235	61.2%
Remain unemployed	116	30.2%
Loss of grazing land	216	56.3%
Family pattern change	123	32.0%
Scarcity of fuel wood	280	72.8%

*multiple answers were counted, Data source: Field survey, 2017

Picture 1: Flood water from Sirajganj district, Data source: Field survey, 2017

In this part of the survey, the respondents were asked to explain the socio-economic problems that they suffered in the last three years due to various natural disasters. A number of sufferings were mentioned by the respondents. For example, in Table 5.3, we can see that 79.2% of respondents claimed that they suffered from a lack of safe drinking water both from floods and droughts. Due to the flood, most of the tube-wells were submerged under the flood water where the villagers were unable to have pure drinking water. A similar scarcity was also experienced due to the drought. The level of water recedes creating difficulties in access of drinking water from the tube-wells. Scarcity of fuel wood was considered as another problem (72.8%) created by natural disasters. Culturally, villagers of Bangladesh in general and villagers of these three villages are used to extracting fuel wood from nature. Due to floods, everything went underwater and villagers could not collect fuel. Traditionally, villagers of Bangladesh are used to eating rice, fish and dal three times a day. For that reason, there is a proverb in Bangla literature, *"mache vat-e Bangali"*. However, the survey data show that 61.2% of respondents claimed they need to change their food habit due to natural disasters. During a flood most of the villagers need to move to a safer place than their own house and leave all materials there. Sometimes they need to stay in a new place for more than a month. During this time, they need to depend on dry foods like biscuits and puffed rice or *chira* which they get as aid from different organizations. Many (30.2%) of them asserted that they need to remain unemployed due to natural disasters like flood. Besides that, 32.0% of respondents avowed that their family pattern had changed due to riverbank ed rosion or land erosion. Many of them needed to move from their own place and leaving behind many of the family members. Moreover, to adjust with the situation, a number of family members were found to conduct migration which was treated as the changing patterns of a family.

Table 5.4: Amount of food production

Food production less than before	Frequency	Percentage (%)
Yes	90	23.4%
No	294	76.6%

Data source: Field survey, 2017

This part of the survey intended to know the effect of natural disasters on agriculture in northern Bangladesh. Moreover, this part may discover the level of food security of the villagers of disaster prone areas. Very surprisingly in Table 5.4, we can see only 23.4% of the respondents decreed that the amount of food production is lesser than any other time of the past. Informal discussions with the respondents assured the reasons of such an answer. As most of the respondents lost their land from river erosion, they do not have their own lands for agriculture; so, most of them do not have any idea on the amount of food production.

Table 5.5: The reasons of less food production than before

Reasons	Frequency	Percentage (%)
Unproductive land	69	76.7%
Drought	49	54.4%
River erosion	66	73.3%
Flood	77	85.6%
Lack of proper irrigation system	18	20.0%
Lack of cattle for plough	14	15.6%
Lack of seed and fertilizer	35	38.9%
Insufficient use of insecticide	26	28.9%

*multiple answers were counted, Data source: Field survey, 2017

Picture 2: Cattles suffering from lack of food during flood. Picture is from Bogra district, Data source: Field survey, 2017

The respondents who claimed of less food production than before, was asked about the reasons of it. A total of 85.6% of the respondents stated in 5.5 that the reason for less food production is the sudden attack of floods. Next to floods, the reason for less food production is the unproductive lands (76.7%). These unproductive lands are the outcome of silt resulted from floods. This part of the survey brought an interesting finding. It revealed that the causes of less food production in these areas are mostly natural, not manmade. Villagers can manage the situation if it is under their control. For example, only 15.6% and 20.0% considered a lack of cattle for plough and a lack of proper irrigation system are the causes of less food production.

Conclusion

A household-based survey with questionnaire was conducted to have the basis for a quantitative characterization of a household of climate change, climate change effects and changes in household resources over the years. A total of 384 households were selected for interviews in three villages from three districts of the Northern part of Bangladesh. The selection procedure was completely random but due to the absence of many households, the random schedule was not followed properly. Similarly, due to the absence of many household heads during the time of interviews, the senior member next to the household heads were interviewed.

The result found that seasonal temperature anomalies are occurring more than before. In recent years, the summer is warmer and longer than the past years and similarly, in winter the unusual cool with shorter term comes around. Due to the weather fluctuations, their livelihood preparations are affected.

CHAPTER 6

Results and Findings: Causes of Migration

The causes and factors of migration

Due to insufficient data, climate change is not a certain issue for many but in general, we can see that the number of cold days decreased and hot days increased in most of the European, Australian and part of the Asian countries (Faist, & Schade, 2013). The ice melting ratio intensifies with increased temperature which causes floods in many parts of the world. Livelihoods of the residents are affected with the changed situation and the created vulnerability pushes the residents to move to a safer place. With environmental disturbs like floods, droughts, desertification and land erosions, the inhabitants cannot adjust and continue their livelihoods anymore and getting no other existing alternative, many of them need to displace. Some of them cross the border while others remain within the border.

The heavy rainfall or heavy droughts resulted from the global climate change shapes the nature of migration in that society. Although climate change and its effects force people to change their location and face challenges, migration creates some opportunities in changing their lives and luck for the betterment. The challenges are mostly for those who are unable or unwilling to migrate. Other than climate change and its effects, the reasons are mostly economic, to join family members, or to avoid social conflict and harassment. These kinds of migration are mostly temporary (Black et al; 2011).

Migration is one of the potential alternatives for individuals and households to reduce the vulnerability and stress from sensitive environmental conditions. A livelihood dependent on agriculture and natural sources choose migration as one of the techniques of adjustment with the changed environment. For example, in the West African Sahel, rural households conduct migration during the period of droughts. During the prolonged dry season, many young men migrate to urban and semi-urban areas for their earnings. Similarly, many households migrate to other rural areas for their earnings. Sometimes, only the young kids are sent to shelter at their relatives' house. This is how the inhabitants of Sahel adjusted themselves with the changed environment (McLeman & Smit, 2006).

Table 6.1: Techniques of adjustment with the entire socio-economic crisis

Techniques of adjustment in entire socio-economic crisis	Frequency	Percentage (%)
Taking loan from NGOs	117	30.5%
Migrating	113	29.5%
Selling property	52	13.6%
Taking loan from relatives	79	20.6%
Waiting for aid from government	106	27.7%
Waiting for aid from non-government organization	109	28.5%

Waiting for aid from public	150	39.2%
Changing of traditional work	137	35.8%

*multiple answers were counted, Data source: Field survey, 2017

In Table 6.1, we can see that there is no single technique that the villagers use most prominently. It was found that a number of alternatives depend on the situation and opportunities that they get to grab. For example, 39.2% of villagers said during a crisis they waited for aid from the public. According to the villagers, before the government and non-government organizations, in most cases they received support from the general people. The general people tried their best to rescue them from hazardous situations and provide the socio-economic support that they have. Likewise, 35.8% of the villagers said they changed their traditional work and looked for alternative work to survive with the crisis. Very expectedly, 30.5% of the villagers were found to be dependent on NGOs to get loans to persist with natural hazards. Interestingly there were similar results for both waiting for aid from government (27.7%) and waiting for aid from non-government organizations (28.5%). Surprisingly, only 13.6% respondents alleged that they sell property to survive the natural disasters. From the connecting informal discussions, it was learnt that most of the villagers do not have property to sell. In most cases they lost their landed property to land erosion. However, 29.5% of respondents uttered that with they prefer to go for migration to look for work during a crisis.

Table 6.2: Techniques of adjusting to a food crisis

Type of technique	Frequency	Percentage (%)
Planting new types of crops	42	11%
Preserving water for agriculture	8	2.1%
Modernizing irrigation infrastructure	22	5.7%
Purchasing all food products from market	375	97.7%

*multiple answers were counted, Data source: Field survey, 2017

They were asked to be very specific about the techniques that they used to overcome their different crisis. In Table 6.2, the techniques in adjusting to a food crisis is shown. Although 11% of the respondents said to increase their food production, they tried to plant new crops; but as most of the villagers lost their land and many of them lived on the government khash lands, they did not have the alternatives to go for agricultural works. Very logically, to adjust with the food crisis, they needed to depend on markets. All the food items they needed to purchase are from markets. Around 97.7% of the villagers said from fish to rice, each and every food item needed to be purchased from others.

Table 6.3: Techniques to adjust with the fuel crisis

Type of technique	Frequency	Percentage (%)
Collect fuel wood by extracting from nature	308	80.4%
Preserving fuel	5	1.3%
Purchasing fuel wood from market	129	33.7%

Data source: Field survey, 2017

In the earlier part of the survey, the villagers mentioned fuel crisis is one of the major sufferings created by disasters. Thus, Table 6.3 depicts the techniques used by the villagers to adapt with the fuel crisis. A significant (80.4%) number of the respondents mentioned that they collect fuel wood by extracting from nature. Informal discussions added that in most cases, female members of the family have the responsibility to extract fuel. They extract by sweeping leaves fallen from the tree. Floods cause huge damage to the leaves. Moreover, a drought also causes scarcity of fuel as scarcity of water does not allow them to grow trees. This is true mostly for Gaibandha district. The crisis of fuel often pushes them to purchase fuel wood from the market to continue their household necessities.

Conclusion

Migration is found to be one of the alternatives among many others to cope with the vulnerabilities resulted from climate change. The other coping mechanisms include assistance from relatives and friends, changing and diversifying livelihood strategies, and support from the credit groups. Even though in terms of times food intake is affected slightly, but the production of agricultural products is affected a lot due to land erosion. On the other hand, during a drought season and mostly in Gaibandha, the crop production is affected enormously by the climate change. The agro field is sandy and not suitable for crop cultivation. So these people are dependent on purchasing from markets for their food supply. In this case, the migration works is one of the reliable source of living and purchasing food from markets.

CHAPTER 7

Results and Findings: Migration and Migrants

Migration and migrants related information

Economic opportunity plays a vital role in individuals' migration. Labor market opportunity is another cause of migration for individuals. To support the urbanization and opportunity to work in the industrial sector, individuals migrate from the less developed part of the country. However, other socio-cultural factors are not less important in initiating the migration process of individuals. Such an explanation was given by MacDonald & MacDonald (1964) where they talked about "chain migration". They uttered that before conducting migration, where the migrants get the information of possible opportunities of a place from the previous migrants. Not only that, after migration the new migrants received some benefits such as accommodation, transportation and even employment opportunities from the previous migrants (MacDonald & MacDonald, 1964). These authors showed the migration tendency of migrants before the First World War. They portray that immigrants of that time were mostly males who moved temporarily for their earnings. Though many of the male migrants were married but their tendency was to leave their family behind. However, the few who bring their families along with them in their first travel, mostly they will do permanent migration in their new place. Other than these migrants, before assimilation with the new society, their intension was to earn money to send back to their family and to build a new house in newly purchased land.

Some other researchers prefer to use the term social networks to explain the chain migration of MacDonald and MacDonald (1964). Social network was considered by many researchers, as one of the vital factors in determining migration decision of a migrant. Poros (2001) specified that interpersonal networks keep a positive role in determining migration to a particular destination and to work in a specific occupation. Black and his associates (2011), on the other hand, stated that there is no single driver for an individual's migration decision, rather, a number of drivers (e.g., economic, political, social and demographic) work together for a migration decision. These drivers are directly affected by the environmental change. For example, with environmental change the economy of the individual or societal is vibrated through the effect on livelihood. Similarly, with an environmental change, a pressure goes on existing resources which creates a political conflict among the inhabitants. This conflict may push the individual for a migration.

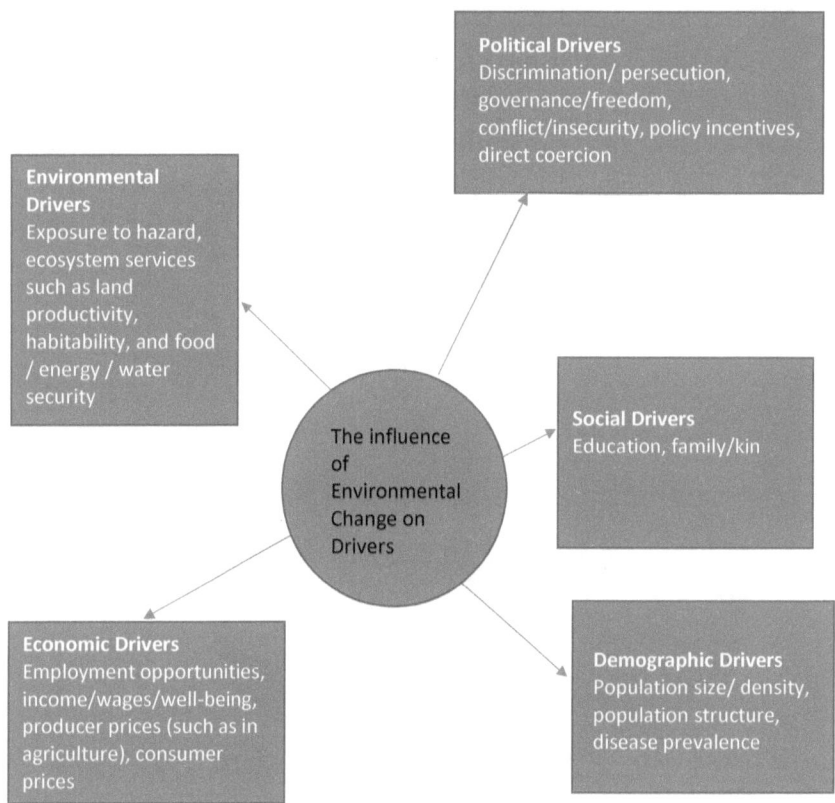

Diagram 7.1: Black, et al; (2011). Slightly modified.

To know the causes and factors that motivate rural people to take part in internal migration to urban areas.

Table 7.1: Number of migrated family members

Number of migrated family members	Frequency	Percentage (%)
1-3	81	21.1%
4-6	2	.5%
7-9	0	0%
12+	0	0%
No member	301	78.4%

Data source: Field survey, 2017

Migration is considered as one of the alternatives by the villagers to adjust with the sufferings created by natural hazards. However, it is found in Table 7.1 that 78.4% of the respondents claimed that till now, no member of their families conducted migration to anywhere. On the other hand, 21.1% of the respondents accepted that at least one member to a maximum of three members of their families have conducted migration to a possible destination.

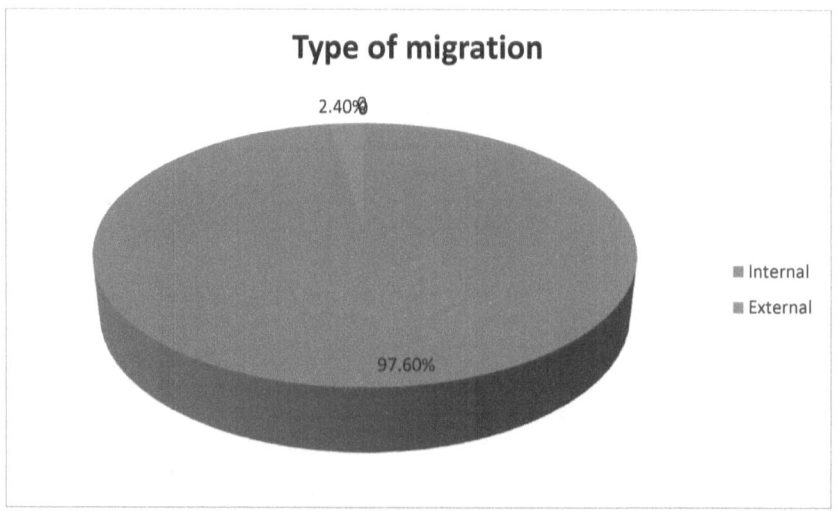

Figure 7.1: Type of Migration, Data source: Field survey, 2017

In this survey, more in-depth information about migration was collected. In the above figure, the type of migration that these vulnerable villagers or their family members conducted is shown. Among the migrated villagers, 97.60% chose for internal (within the country) migration, whereas only 2.40% of them went for external (beyond the country) migration.

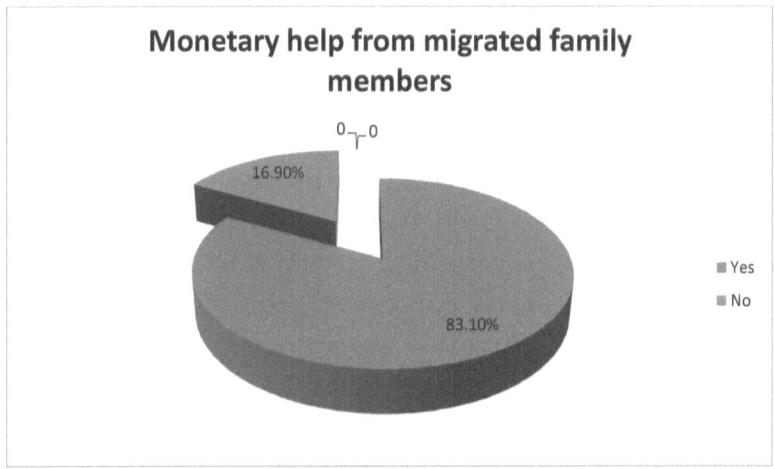

Figure 7.2: Monetary help from migrated family members, Data source: Field survey, 2017

In one part of the survey, we found that villagers needed to purchase almost all the food items from markets. So, money is considered as one of the most important asset to survive. From the migration and related information shown in the above figure, more than 80% (83.10%) of the respondents received financial assistance from their migrated family members.

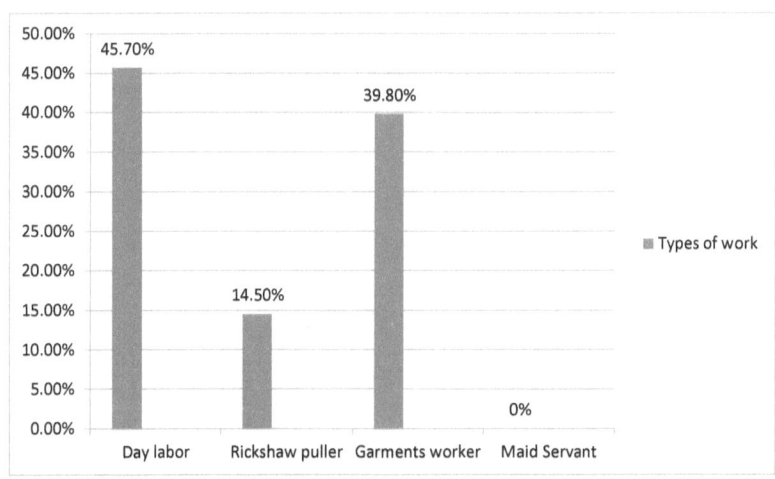

Figure 7.3: Types of work done by the migrated family members, Data source: Field survey, 2017

In the above figure, the information about the types of work done by the migrated family members is shown. It is known that the migrated people mostly (45.70%) do work as a day laborer. This profession includes labor working for agriculture in other villages, working in mud logging projects, road development projects or working as construction labor. Next to labor, migrated people work in the garment sectors as garments workers (39.80%). This survey did not get any information about migrated people who are engaged as maid servants to do household works with payment. The number of migrated individuals as rickshaw pullers (14.50%) is not insignificant. They have also contribution to support their families.

Table 7.2: Amount of money (per month)

Amount of money (per month)	Frequency	Percentage (%)
Less than 500	11	14.3%
500–900	18	23.4%
1000–1900	12	15.6%
2000–2900	14	18.2%
3000–3900	10	12.99%
4000–4900	6	7.8%
5000–5900	4	5.2%
More than 6000	1	1.3%

Data source: Field survey, 2017

In Table 7.2, we can see that per month the amount of money that the migrated family members send to their relatives is very negligible. From this, only 14.3% respondents acknowledged they get financial support from their migrated member more than TK 400/ (per month). On the other hand, more than 80% (85.7%) respondents are receiving less than TK 4000/.So very logically we can see the relationship of occupation of the migrated people and this tends to help them out.

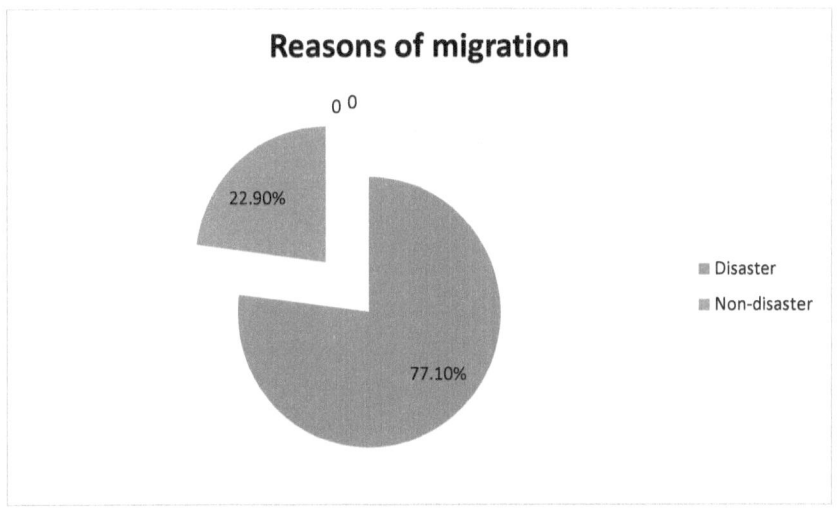

Figure 7.4: Reasons for their migration, Data source: Field survey, 2017

Due to climate change, villagers are suffering enormously from floods, droughts, heavy rain fall, river bank erosion and desertification throughout the year. Since these disasters do not allow them to breathe comfortably in the village, they look for alternatives to survive. Thus the villagers mentioned "disasters" as the reason for their migration to other places.

Table 7.3: Relationship with the migrated member

Relationship with the migrated member	Frequency	Percentage (%)
Son	64	77.11%
Daughter	7	8.4%
Son in law	5	6.0%
Daughter in law	2	2.4%
Grandson	1	1.2%
Other	4	4.8%

Data source: Field survey, 2017

In Table 7.3, we can see that in most cases the son of the household migrated. Socio-culturally and from pre-historic perspectives, sons are considered the asset for the family. Though after modernization, this mentality has changed slightly, but still parents of the village considered their sons as their asset. Since sons take responsibility for arranging the livelihood for their family, they are taught so. Moreover, the social security in the new place is generally unknown. So the family does not allow the female members to migrate for the sake of the family. Since 77.1% of the migrated members are the son of the respondents, this reality came into true in this survey also.

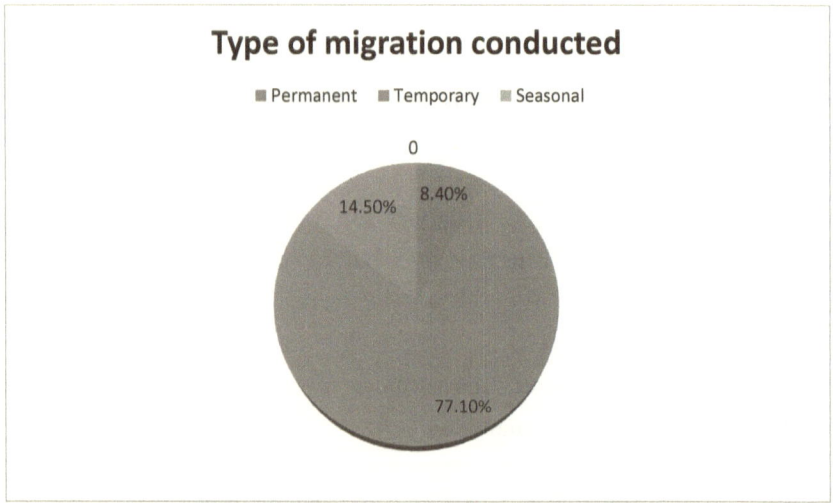

Figure 7.5: Type of migration conducted by family members, Data source: Field survey, 2017

In Figure 7.5 we can see that villagers usually prefer to go for temporary (77.10%) migration. In some cases they chose for seasonal (14.50%) migration also. In some seasons, especially during the paddy harvesting time, villagers move to other places for work. Similarly, with the hope to change their fortune, they conduct temporary migration.

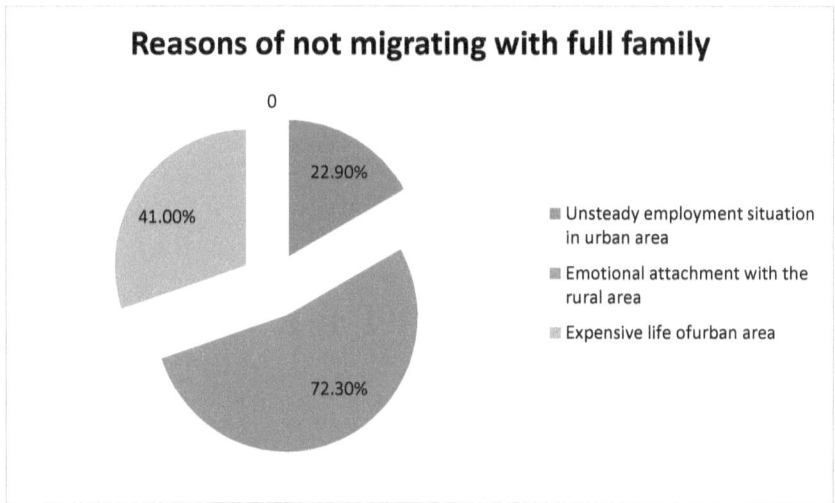

Figure 7.6: Reasons of not migrating with full family, Data source: Field survey, 2017

Field data shows in the figure that migrated people move to another place alone and keep their family behind in their own village. Therefore, the respondents were asked the reasons of not following their relatives to the new place. In this regard, 72.3% respondents stated about their emotional attachment with the village. Consequently, 41% of them also mentioned about the expensive life of urban area which they scared extremely.

CHAPTER 8

Results and Findings: Migrants Experience

Introduction

Bangladesh is frequently considered as a country of natural disasters and in direct threat from climate change. The reason of this sensitive vulnerability is the geographical location of this country as Humalayas is in its North and The Bay of Bengal in its South. This atypical geographical position of this country made the country vulnerable to climate change and its effects (Karim et al; 2017). The temperature of this North-west region increased by 5° over the last 100 years which is calculated by the regional weather station of Bogra (Coirolo & Rahman, 2014). However, in a trend analysis, Bhuyan and his associates claimed that the average rainfall decreased at a rate of 8.8 mm and 40.1 mm in the last 100 years (Bhuyan, Islam & Bhuiyan, 2018). Water layer recedes and the increased demand of water for irrigation and domestic purposes cause droughts in these areas. Similarly, by some research (e.g., Shahid, 2011) it is predicted that in Bangladesh by 2030, the amount of rainfall will increase by 5% to 6%. It is also said that around 72% of its total rainfall occurs only in the rainy season. Although the people of Bangladesh have a very low contribution to climate change, but its geographical location, economic condition, faulty policies, lack of awareness and effects of global climate change, make the people of Bangladesh vulnerable to climate change. In this following section, the details of migrants' experience as a measure to avoid vulnerability against climate change is documented.

Case Study 1: Md. Abdul Mannan, Shaghata, Gaibandha

I am Md. Abdul Mannan. I am 46 years old. My family is of 5 members where I have two daughters and one son with their mother. My elder daughter got married but the bride wants dowry from me. As I could not manage to give dowry, he never takes my daughter to his house. Now I migrated to Dhaka with all my family members. I had two rooms up the embankment my home is in Shahghata Thana Road, Gaibandha. My family is originally from Haisalkandi village of Shaghata. We had two *bigha* of land which we sold out because of poverty. After selling all lands of Haisalkandi, around 8 years before we shifted to governmental barrage of Shaghata. Here many people get support from Chairman and Member of local government but everyone does not get support from them equally. I am one of the unfortunate who do not have good network with these political persons so my life was very vulnerable in my village. **The productivity of our area is very low. We could not go for agricultural work. Our land is full of sand. Everywhere around our area is only sand. This is because of Tista Barrage. However during rainy season when the barrage is opened, our area becomes flooded. So we cannot cultivate any crops** except few pulses. In this season, landless people like us cannot think of agricultural works in our area.** As I do not have any education, what I can do to support my family except doing some low level manual work. In Shaghata, my occupation was to pull tricycle. However, seasonally I did business also to support my family. I used to sell rice in the market of Shaghata. To meet expense of my daughter's marriage and other family expense, I lost all my capital to continue my business. Moreover income in Shaghata was 200/300 per day which was very low to maintain the whole family. So two or three years back, I alone came to Dhaka to change my luck. I stayed in Dhaka around six months and came back to Shaghata. Poverty did not leave me, so this time I brought my whole family here in Dhaka. **One of my friend whom I met in Dhaka helped me to get a job in Dhaka. Currently, I am working as a security guard in an apartment**. Now my monthly income is around Taka 7000. I have rented a two room house which cost 3000 taka per month. My wife is working as a maid in nearby house where she is getting TK 6000 per months. My elder daughter and son are working in garments. Their earning is also more than taka 3000 per months. My elder daughter helped me by giving money but son never wish to give money. However, this is true that in Shaghata we never could be able to take our three times meals but here we are able to eat three times in a day. **I do not have a plan to stay in this house long time. I just want to settle all my kids in work and to save some capital to start a small business in my area again.**

Analysis Case 1:

For Md. Abdul Mannan, a middle-aged (46 years) man of Gaibandha, it is found how vulnerable his life was due to the impact of climate change. In his area (Shaghata, Gaibandha), they mostly faced two major natural disasters which are floods and droughts.

These disasters come in their life by rotation throughout the year. These two disasters come with a number of impacts which cause vulnerability in their daily life. Md. Abdul Mannan mentioned that due to the droughts and floods, the lands of their areas were barren and unproductive. In Gaibanda, in accordance with Md. Abdul Mannan, they faced a huge scarcity of drinking water throughout the year as the water level goes down during a drought. Although Md. Abdul Mannan is not an educated person, he pointed international politics between India and Bangladesh in sharing a common river water as one of the reasons for their sufferings from climate change. Climate change, through the natural disasters had negative impacts on Md. Abdul Mannan's life. He remained unemployed several times of the year. Though he did a few informal works over there but that was insufficient in running a big family of five members. His family life was even tougher as his elder daughter's in-laws refused to take her back for failing to give them dowry. Not only was the dowry, his life in Gaibandha was vulnerable because he didn't have a good network with the political and social leaders. These leaders are powerful in giving someone work and other social support. According to Md. Abdul Mannan, with no other alternatives in place, he was bound to migrated to Dhaka. In the beginning, he moved to Dhaka alone while his family remained in Gaibandha. He used to stay in Dhaka for a few months and then visited his family with some savings. However, he was unable to maintain his family expenses with only his single income. So his poverty pushed him to move to Dhaka with his whole family. To move to Dhaka, his social networks helped him to get a job in Dhaka city. Now his family is enjoying a better life in Dhaka as almost all the family members are working in different sectors and contributing more or less for the family. Even though this family is enjoying a well-off life in Dhaka, Md. Abdul Mannan still dreams to go back to his place of origin. After establishing all his children in different working sectors and having some savings for the future, he plans to go back to his own place.

Case Study 2: Manju Sarker, Sariakandi, Bogra

I am Manju Sarker. My father Ajhar Sarker is the inhabitants of Boyrakandi village of Kutubpur Union of Sariakandi, Bogra. Reality is our full family migrated to Boyrakandi village from Ponnyibari village. Our Ponnyibari village is now in the grave of river.

In my family, we have six members, including my parents and two siblings. Two years before, I have moved to Dhaka for the seek of work. Before losing our all properties in riverbank erosion, I did study in my village. At that time we had around 2 bigha of lands for agriculture and 29 decimals of land for own house. I was used to help my father in his agricultural work. We were habituated to produce paddy, jute, wheat in our fields. But we lost everything within few years of time. We lost our cultivable lands and our residential land also. We had no other choice to move to Boyrakandi. We do not have enough money to buy land here in Boyrakandi. So we took shelter in government land on the roads. Even we do not have plan to buy land here as this village is also vulnerable to be eroted by the river anytime.

After moving to Boyrapukur, my father became paralized and then all responsibilities went to me. So i migrated to Dhaka for work. Yes, in my area we have some opportunities to work but the payment is so less to maintain my family. So one of my cousin brother helped me to get work in Dhaka. Now I am working as a day labor in construction sector. Here i am getting payment in daily basis. I am getting Taka 300-400 per day but the work is very uncertain. If I get sick or the weather is not suitable to work, I do not get any money. Thus my income varies from one month to another. In Dhaka I live with my co-workers in the same place of my work. So I do not need any cost for renting house. After keeping my own expenses, I try to send at least Taka 5000 to my family.

> With this money, two younger siblings of mine are continuing their studies. Moreover, my father's treatment expenses also comes from it. Social networks help us a lot to get related works one after another. So I am planning to stay few more years in Dhaka and if possible after saving some money I will go back to my family again.

Analysis case 2:

For the case of Manju Sarker, of Sariakandi, Bogra, like Md. Abdul Mannan of Gaibandha, a similar scenario was revealed. His family also shifted to the current village due to the climate change and its effects. As their place of origin, Ponnyibari village went to the grave of the river due to river erosion; his family had no alternative to look for a new shelter (Boyrakandi). In Ponnyibari, his family was engaged in agriculture and agriculture-related works. However, all their lands were washed out through the river erosion. In the new place, as they do not have the capacity to buy agricultural and residential lands, they took shelter in the government's plot. Being the elder son of a large family, Manju Sarker faced hardship in maintaining the family expenses as he had no land or work opportunity which pushed him to migrate to Dhaka. Not only these issues, other factors like sickness of his father and large dependent family also pushed him to migrate to Dhaka. To migrate and work in Dhaka, his social network like his male cousin helped him. This temporary migration of Manju Sarker helped him in sending money to his family on a regular basis for his younger sibling's education and to continue his father's treatment. Although his income in Dhaka city was not certain and persistent, his life in Sariakandi, Bogra was even more uncertain.

Case Study 3: NurAlam. Sariakandi, Bogra.

I am NurAlam. I am 17 years old. This year I sat for Dhakil Exam (equivalent to SSC exam) and now waiting for the result to be published. We are three members in my family. In my family, I have my mother and an elder brother. **Like many others, we migrated to Boyrakandi village from Punnyibari. We lost everything through river erosion. I have no memory of Punniybari, as I born in Boyrakandi.** My father passed away when I was a small baby. We never had enough money to buy land for our own residence. So we are living on the embankments that the government built to protect the village from river erosion. Later my elder brother migrated to Dhaka for work as we lost our father long before. My brother is working in a garment of Shafipur, Dhaka. Every month, he sends us Taka 3000 for me and my mother. However this is not enough to maintain our family as we need to buy everything from the markets. We have a net which I use to catch fish from the river. My brother has a dream to make me an engineer. For that I do study properly. However it is quite difficult to buy books and other required stationaries for my study. As I get a break after the exam, to earn my study expense, I came to Dhaka last week. This is not first time. Before this, I came in Dhaka one more time. That time after doing registration in Grade 9, I was needed to buy new books of around Taka 3000. My family was unable to pay that extra money. So I came to Dhaka for one month. I did work in building construction. One of my uncles helped me to get that work. **I earned more than taka 3000 in one month and went back to my village again. Now I came here taking that uncle's help again. I started to work in building construction for a second time. I heard as I am amateur construction labor, the owner will give me less payment for my work.** Per day, they will pay me taka 100-200. I will work here for around 2months and will go back again to start my study in the village.

> **We do not have our own tube-well but we collect our drinking water from my uncle's house. They helped us with best of their capacity.** Last year we were able to set up a latrine for our own. Before that we were used to use open place. My brother with his hard work managed to set up a latrine. **Beside this, last year we took a lease of a calf. We feed it grass and some organic foods that are locally available. Nowadays, grasses are not available like before. My mother needs to collect the grass from the *char*. She takes help from our uncle to go to the *char* by boat. She collected the grass for at least 2 days at once.** We are planning to sell the cow in the coming Eid-ul Adha to have a good price which we may need to share the original owner of the cow. After getting admission in grade 11, I will come back again in Dhaka to earn my academic expense.

Analysis Case 3:

Nur Alam, a young man of 17 years old, also conducted a temporary migration to Dhaka city from Sariakandi, Bogra. Like many others, his family shifted primarily to Boyrakandi village from Ponnyibari as they lost all their assets in river erosion. Nur Alam is an orphan with a big dream of becoming an engineer in the future. But to fulfill his dream, he needs to face a number of obstacles in the society. To maintain the family expenses, his elder brother had migrated to Dhaka a couple of years ago but the money that he used to send back home was not enough. So both Nur Alam and his mother needed to do additional work.

> **Case study 4: Aslam Mia, Meghai, Sirajganj.**
>
> I am Aslam Mia. I am 26 years old. I came in Dhaka 3 years before. **I started to work in a garment and together with that I continued my study in degree level. I need to move to Dhaka as we have no property in Meghai to survive.** Moreover, shortage of works pushed me to move to Dhaka. **Our full family migrated to Meghai in 1988 as we lost our house of devastating flood of '88. We had lands for agriculture but I have no idea about the amount of it. My grandmother did not get chance to hand it over to my parents as she lost all lands and papers as well during '88 flood.** I got married last year. My monthly income is TK 10,000 which is not sufficient to maintain a family in Dhaka. So **I could not manage money to pay to my family of Meghai. My father works there as a day labor in agriculture sector**. But every year, I visit my family. **In our house, we have both the tube-well and sanitary latrine. But both in flood and drought time, our family suffer from fresh drinking water. During that crisis time we need to collect drinking water from other's house.** I will not stay in Dhaka long time. I will go back to my place after managing some capital to start a new business in my village.

Analysis of Case Study 4:

Aslam is a married young man who migrated to Dhaka city to change his and his family's life. His family lost all properties including agricultural and household lands in a devastating flood in 1988. To support his family he migrated to Dhaka from Meghai, Kazipur, Sirajganj and worked in Garments but the wage in the garments sector was so little that he was unable to support his family residing in the village. At the early stage of his migration, he tried to provide financial help to his parents but after getting married he was no longer able with his little income. So, in the village his old father needed to work to continue his life. Though Aslam set a tube-well for his parents but during the floods and droughts of every year, they were not able to have drinking water from this tube-well. His old mother needed to walk a long way to get pure water; and with no alternative, sometimes they drank impure water from the flood. Similarly, although Aslam set a toilet for his parents but this went under the flood water and remained unusable for a couple of months. During these days his parents needed to use an open place which was even more scarce during a flood. Like others, he also wants to go back to his place of origin and to settle down with a new business over there.

Case Study 5: Sajedul Islam, 32 years, Sariakandi, Bogra.

I am Sajedul Islam. I am 32 years old. My family lives in Boyrakandi village of Sariakandi, Bogra. Before that we lived in Punnyipukur village. **We had around 20 *bigha* of lands in Punnyipukur. We lost it gradually from every year's attack of floods**. Our previous village Punnyipukur is absolutely in the grave of the river. Finally we were needed to move to Boyrakandi. We do not have land for residence in Boyrakandi. We live on the government's embankment built to protect the village from riverbank erosion.

We are seven members in our family. With my parents, I have two sisters, my wife and a baby in my family. **In my previous village, I was used to do agricultural work. Since we lost our lands in river, we became workless. So I need to move to Dhaka for the sustenance of my family.** My father is sick. So he cannot do any work to help the household. So after coming in Boyrapukur, I had no other choice to migrate to Dhaka for work. I have migrated to Dhaka alone, keeping my whole family in Boyrakandi.

As I am working in Dhaka for more than seven years, the owner pays me higher than many amateur workers. **I am a day labor in construction sector. Seven to 8 years before, one of our uncles like neighbor offered me to come for work.** He just helped me; he never charged any money for that. After that, **a network established with the coworkers. Every 3-4 months, I visit my family to Sariakandi. Per day, my owner pays me around TK 400 cash for my work.** After keeping my own expenses, I can pay my family around 7 thousand taka. My family is a big family. It is not so easy to maintain the family with the limited earnings. Moreover, my two sisters are doing study. As we need to buy each and everything from the market, it is not easy to maintain the household expense. **We suffer badly from scarcity of drinking water crisis especially during flood time. All the tube well becomes flooded with the water. In every year at least 15 days to one month, we had to suffer from it.**

If the work is finished in one place, it takes time to get another work. In that break, I go back to my village. After getting a new work, I usually come back again. The work I do, I sometimes very uncertain as the owner pays money with his own choice. But this is true that if I do not come here in Dhaka, our full family may suffer badly. **As life is very expensive in Dhaka, I cannot be very brave to bring my family here.**

Analysis of Case Study 5:

To maintain the responsibility of a big family, Sajedul Islam of Sariakandi, Bogra needed to migrate temporarily to Dhaka. As he had no formal education, he worked as a day labor in the construction sector. With his limited income, he tried to support his family in the village with their household expenses. Not only that, with his limited income he tried to support his sisters' education. Sajedul and his family were middle class farmers in the Punnyipukur village with more than

20 *bigha* of agricultural lands which they lost gradually with every year's attack of floods. At present they do not have a single piece of agro land. After losing all the lands to the river, they moved to another village named Boyrakandi in the same sub-district of Bogra. His good network with his neighbor helped him to get work in Dhaka city at the beginning and now he has established a good network with his co-workers. So, once he finishes his work in any project, with his networking he can manage another work in another place, but in between he stayed jobless as his job was not fixed for anyone. Sajedul needed to maintain a big family with his little wage, so he was not able to set any tube-well or latrine for his family. His family suffered due to the scarcity of drinking water during both the flood and drought period. The women of his family walked a long way in search of drinking water.

Case Study 6: Shahidul Islam, Sariakandi, Bogra

I am Shahidul Islam of 55 years old. My education is till primary level. My family resides in Boyrakandi village but my origin is in Ponnyibari village. This Ponnyibari village does not exist anymore due to river erosion. **In Ponnyibari, we had around 7 *bigha* of lands which we were used to use for cultivation. Due to river erosion, all our lands gradually eroded completely in Ponnyibari.** We had to face poverty having no alternative. So around seven years before I had to migrate to Dhaka for earning our living. However we need to shift to Boyrakandi from Ponnyibari once we lost our residential land in river erosion. **I have two sons and their mother in my family. Both of my sons are doing study.** The elder one is studying in grade 12 and the younger one is in high school. **It is very difficult to get work in our village. Most of the work is very low paid so for the survival I have no other choice to migrate in the urban. In Dhaka, I do work in house building sector as a day labor.** Per day, my income is TK 350. After keeping my own expenses, I am able to pay 4000-5000 taka per month to my family.

> Every 3 to 4 month, I tried to visit my family. It is quite expensive to go even after 3 to 4 months. It takes almost a month income to visit once. Since there is no work even in Dhaka during the rainy season, I had no choice but to come back to my village. Beside rainy season, even in other times, we cannot do our work; owner of the project does not give us any money. We need to do struggle at that time. **People of our own areas helped me to come in Dhaka to get work. Now I established networking even with other people who help me to get work if I became workless.** As I already become aged, I do not feel comfortable to work here leaving my family. I am waiting for my sons to start their working life so that I can go for retirement.

Analysis Case Study 6:

Sahidul Islam, a middle-aged, migrated day labor from Shariakandi, Bogra to Dhaka, worked in the construction sector after losing all his agricultural land in a river erosion. Due to his age, he could not work every day; thus his income was less than many of his coworkers. Around seven years earlier, he migrated to Dhaka with the help of one of his neighbors. When one project is completed, he got work in another project very easily because he had already built a network with many people who can arrange work for him. However, during the rainy season he remained workless as most of the construction projects paused their work in this time. With his limited income, he tried to support his two sons in their studies. As there was very limited options of working in his village, he was waiting for his sons to become mature and to start their working life in Dhaka or any other city so that he retire in peace.

Case Study 7: Tofajjol Islam. Sariakandi. Bogra

My name is Tofajjol Islam. I am 50 years old. I have two sons. We are four in total in my family. **I did study till the Higher Secondary Certificates (HSC). After that I could not continue my study due to poverty.** My village was in Ponnyibari. **In every year we lost land through river erosion. When we lost even our place of residence, we were bound to shift to the nearby village which is Boyrakandi.**

From last 12 years, my family is living in Boyrakandi up the embankment. Though every year this place also faces floods but till now we can stay here. Once we had everything but now we do not have even a single decimal of lands. In our village and **nearby areas, we do not have work. Actually most of the villagers are facing land eroding problems, and then who will provide us work!!!** So around 13 years back, before shifting in Boyrakandi, I came for work in Dhaka.

I do not stay here continuously. After coming here 13 years before, I worked here for around 1 year. I stayed in village for some years and finally 3 years before I came back in Dhaka once again. **This time one of my uncles helped me to get the work.**

Per month, I can earn around tk 7000. After keeping my own expenses, **I can send around 5000/ to my family. The reality is I cannot assure my income. It depends mostly on the working day.** If it rains heavily or the work is stopped for any uncontrolled reason, I do not get payment. Per month income fluctuates. **I have set up a tube well in my own place. So collecting water is reduced a lot than before. As during floods and droughts the water layer is changed, so we face difficulties to get pure water.** We have cow and goat as well. If my family faces any crisis of money, we need to sell them off to meet up the crisis. My sons are growing and I am becoming old. So once my sons start their works in a suitable place, I will go for retirement and will go back to Boyrakandi.

Analysis of Case Study 7:

Like many other villagers, this middle-aged gentle man faced multiple migrations during his lifetime. As long as this man and his family had land of their own, they tried to stay in their own village. After losing all properties in river erosion, he needed to move to the town for work to support his family. As these people like him were with a poor education level and they could not manage formal work. Therefore, their work were very uncertain in a formal sector. However, their social networks with their friends and family helped them to get new jobs in their time of crisis. He sent back the maximum amount from his earnings as he has school-going kids in his family who required money for their education. Even though his family had a tube well for drinking water, during rainy and summer seasons they were unable to use this tube well. Either they had to drink available unhygienic water or go a long way to other villages to collect pure drinking water. Like many others in his situation, he had a plan to return to his village once his kids are able to earn to support him.

Case Study 8: Abu Sayeed, Meghai. Sirajganj

I am 30 years old. After passing SSC, I came to Dhaka 10 years before. We live in New Meghai but our original village was opposite of the river. Name of that village is old Meghai. We lost all our lands in river erosion. We were needed to move to new Meghai at that time to take shelter. Till now we are living in new Meghai. After coming in new Meghai as we lost all our means to survive. So **I need to come to Dhaka for our survival. I started to work in a garment factory. Per month I am getting around 9000/.** As I lived in Dhaka with my wife and daughter, I feel really hardship to maintain. However **keeping my monthly expense, every month I try to send 3000/ to my parents who are residing in Meghai.**

> My parents stay with my elder brother and his family. We purchase 8 decimals land for our residence. We do have our own tube-well and latrine. Life is very destitute in Dhaka. **I do not want to stay here long time but in Meghai our life is surrendered to the nature. I dreamt to go back to Meghai to start a small business over there.**

Analysis of Case Study 8:

After losing all lands in a river erosion and getting no means to survive, Abu Sayeed had to migrate to Dhaka at a very young age which disrupted his education. Primarily, Sayeed's family moved to another village to take shelter from the natural disasters. However, later on Sayeed moved to Dhaka to work in a garments factory. With a small wage, although he struggled to maintain his family with a wife and daughter in Dhaka, he still tried to support his parents living in Meghai, Sirajganj. With his small support, his parent were able to purchase land for their residence and to set a tube well and latrine in their house. He was stressed in the metro life of Dhaka and desired to return to his village with some capital to start his business in Sirajganj.

> **Case Study 9: Latif, Meghai, Sirajganj.**
>
> I am 35 years old. **I did study up to primary school. I am working in a garment factory of Dhaka. I am living in Dhaka with my wife**. Our house is on the embankment of Meghai, Sirajganj. My parents and younger brother live over there. I have two younger sisters as well but they got married and stay with their in laws. My younger brother is continuing his study. We had some pieces of lands which are smashed with the wave of river. To support the family, I was needed to migrate to Dhaka. **Now my monthly earning is around 10000/.**

> **Every month I send 4000/ for my parents.** Our family made two tin shade rooms on the embankments. My parents stay in that house with very hardship. It will not possible to bring my wife in that house. **I want to build a brick house. After that I will go back to my village**. However every year I visit my parents whenever I get chance.

Analysis of Case Study 9:

After losing all lands in a river erosion, Latif migrated to Dhaka and quit his education at a very early age. Later on, he got a job in the garments sector with very little wage. With this wage he faced hardship to maintain two families; one in Dhaka and the other in Sirajganj. With his financial support, his old parents could survive. With his support, his younger brother could continue his study till today. As his education level was very poor, he was not able to earn more and to further support his parents. Although he made two tin shaded rooms in his village, he was not able to build any toilet or kitchen in his village. He was not able to set any tube well in his yard. In his village, he was not even able to provide any support for livestock or poultry. As his family lived in the government embankment and as they did not have home yard, they were not able do gardening for vegetation. Latif is such an unfortunate migrant that due to poverty and lack of shelter, he was not able to bring his wife and child in the village. However, like many other migrants, he dreams to go back to his village permanently after building a concrete house over there.

> **Case Study 10: Farid, Meghai, Kazipur, Sirajganj**
>
> I am 35 years old. My family lives in Meghai. I have my parents, two younger brothers and a sister over there. Beside them, I have one sister who got married last year. My brothers are studying. For their safety, as elder brother I was needed to migrate to Dhaka city 12 years before.

> **After passing my HSC, I migrated to Dhaka. I am a small entrepreneur. I have a small business of a grocery shop.** With my own money, I completed my bachelor degree also. **Now I try to pay at least 2000/ for my brother's study. I have set up a tube-well and a latrine in our house. I heard, every year, water layer of the tube-well goes down during the summer.** We never had much land to survive. **My father was a day labor in agriculture. He became old and cannot work anymore. My family depends on my support.** I do not have any plan to go back to my place again. The business I am running here in Dhaka will not continue in Meghai.

Analysis of case study 10:

Unlike many others, Farid of Meghai, Sirajganj had no plans to return to his place again. He had instead conducted permanent migration to Dhaka city. At the beginning of his migration, he did not know the types of migration; his primary intention at that time was only to support his family. However, later on, as he got better options in earning in Dhaka, he preferred not to go back to the village of uncertainty once again. Unlike others, Farid got a better option of earning in Dhaka city because he is a graduate. After his SSC examination, he moved to Dhaka from Sirajganj. He started to work in the garments sector at that time with a small wage. With his small wage, he did not stop his education. He became a graduate and looked for an independent entrepreneurship. Now with a small grocery shop, he helped his family of Meghai on a regular basis and his family was able to live on his earnings. After losing their household to a river erosion, his family resided on government land. His father was a day laborer in agriculture but now very few farmers in their village have agriculture work throughout the year. So his father remained jobless for most of the time in the year. Moreover, being an old man, he was not able to work much to support the family. So being the elder son of the family, it was Farid's responsibility to help his family. This successful entrepreneur was happy with his current position as he was able to send money for his younger brother's education also. Moreover, he set a tube well for his parents in his village. However, his concern of not getting drinking water in the summer season was high. Nevertheless, he was happy that he was able to set a latrine for his family and they did not have to use the open place if needed.

Case Study 11: Aslam Mia, Meghai, Kazipur, Sirajganj.

I am 26 years old. I am studying in bachelor degree and at a time working in a garment factory. I have migrated to Dhaka around 5 years before. I heard our family lost all the agricultural lands in the devastating flood of 1988. Now we do not have any agricultural field. However my grandmother has a piece of commercial land in market place of meghai. As she did not pass this land to my father yet, our family cannot use it. **In meghai, I had pulled rickshaw van. Sometimes I worked as a CNG driver. If needed I did agriculture related works in others field with payment. Due to poverty, I was bound to migrate in Dhaka.** Now I got married and living in Dhaka with my wife. **Though I can earn around 10000/ in every month, but I cannot send single penny to my parents. Dhaka's life is very expensive**. We need to struggle a lot with this limited income. So I cannot help my parents financially. My parents live in government's *khash* land which was primarily allotted for government hospital. **Though in Meghai we have a house but we do not have our own toilet or tube-well. We need to collect water from going other's place.** My family lives in poverty but I cannot help them for my limited income. I want to go back to my family. I want to utilize my grandmother's land for my business.

Analysis of case study 11:

Aslam Mia, a young man of 26, started his working life at a very early age. To reduce the acuteness of family poverty he did diversified works based on the availability. This man sometimes worked as a tricycle puller or sometimes as a CNG driver. For earnings, he worked in the agricultural sector too as a day laborer. Although he and his family suffered a lot due to poverty, he did not quit his education. After migrating to Dhaka city for better earnings, he did not give up his studies. He worked in the garments sector and continued his studies. His current monthly earning was very minimal and with this income he was unable to support his family of Sirajganj. He dreamt to have a better job after completing his graduation. Just after his migration to Dhaka, for a few months he was able to provide financial support to his parents. But as his salary did not increase and he got married with that limited income, he was unable to send financial support to his parents. Like some other migrants, his family also lived in the government *khash* land as they lost all their lands in a river erosion. His family took shelter in this *khash* land after losing all their properties in the 1988 devastating flood. After that flood, this family was not resilient anymore. During the summer and rainy seasons, this family suffered a lot as they did not have any personal toilet and tube well. Female members of this family walked a long way to collect drinking water during the summer. This family used open unhealthy planetary for toilet purposes. Although the members migrated to other places for work, this family was unable to get back their resiliency as they had more dependent members than earning members. This family had no source of other earnings such as poultry or livestock.

> **Case Study 12: Asadul, Meghai, Kazipur, Sirajganj.**
>
> I am 28 years old. I am married and I studied till the 9th grade. **We had around 5 bigha lands which we lost in river erosion. At my village, I was jobless. I had no work to earn. So I have migrated to Rajshahi around 7 years before to earn and to support my family.** In my family, I have my parents, wife and a son. They all live in Meghai. I am working in private bus service as a contractor. **Monthly, my income is around 7000/ from where I manage to send 4000/to my family.** Though we were able to set up a tube-well but we do not have our own toilet. So **we need to share toilet with other families and in some cases we need to use the open places. I want to stay here till I can solve our entire family problem.** After that I will go back to my village. I have a plan to start my business.

Analysis case study 12:

Divisional cities, like Rajshahi is also a destination for migration. This young man left his village after losing their agricultural lands in a river erosion. As he left his school at a very early age, he could only manage to work in an informal sector. From his low income he managed to give a part to his family residing in the village. But with this income he could not afford to build a hygienic toilet of their own. Like other migrants, this young man also desired to return to his village after solving all his family needs.

Conclusion

Floods, droughts, *monga*, and river erosion are some of the challenging realities for the people of this region. The alleviation from these challenges depends on the socio-political condition of these people. The awareness and skill development programmes and policies are not executed in this region for these people. Therefore, their dependency on conducting migration as per their ability works as the most deserving solution to come out from the vulnerability. However, it is a matter of concern that this migration trend should be monitored properly by the authority. Otherwise, if the pattern of migration continues haphazardly in this way, it may cause other kinds of socio-environmental problems in the new workplaces. The migrants are paid a very low wage with no job security, and live in a cheap and low housing system with a very filthy and messy surrounding which put these migrants in an unhealthy condition. Although these migrants are contributing greatly in uplifting the economy of Bangladesh, there are no policies and guidelines for them regarding their housing, health care facilities, data centre with job information etc. If these issues can be incorporated with the existing policies, these migrants can be the true resource of Bangladesh.

CHAPTER 9

Results and Findings: Experiences of Household of Origin

Introduction

Climate change triggers the people of less developed countries for migration (Meze-Hausken, 2000). Two characteristics were found very noticeable in these three villages. The first characteristic is that these villages have a high exposure to environmental extremities and the villagers have a long tradition of mobility. Accordingly, there are a number of evidences found of community replacement following natural disasters like floods, river erosion or drought. The primary movement of the villagers are at the community level. The entire village shift to nearer available places. The secondary movement of the villagers are mostly internal and to the town for non-farming work. Most internal migrants in these villages could be measured as the economic migrants, moving from rural areas to urban areas in search of work. In some cases, rural to rural migration to support the rural labor market in agriculture and other non-farming informal sectors are also evident. Lack of work and food security make these migrations as forced migration. However, no evidence is found where their migration type can be termed as refugee as none of them left their place due to fear or anxiety.

The migrants of these villages move in to nearer towns or to the capital city of Dhaka. The poorest of these poor cannot migrate due to their lack of information of where to migrate, how to migrate and what to do after migration. The poor who are able to migrate send back money to their relatives residing in their own village. Even though the amount of money is small, but this amount contributes to their household income and capital.

Other than the psychological loss of keeping the beloved one in the distance, apparently the contributions of migration are positive for the household adaptation to climate change. The economic support that the migrants create for their relatives has multiple constructive effects on their life. These relatives are found comfortable with food security throughout the year compared to other groups. These households with migrant members are found in better condition during the time of crisis. However, the adaptability of these households mostly varies based on the migrants' duration of migration, their type of work, per month income, number of dependent members in the place of origin, and the amount of money sent by the migrants.

Case study 1: Md. Rabbil Ajij Talukdar. Meghai, Kazipur. Sirajganj.

I have passed SSC exam. I have my wife, four sons and a daughter in my family. Among these my daughter already got married and **two elder sons migrated to Dhaka to earn their living.** The younger two sons are living with me in the village. Though I am living in Meghai village but this is not my village of origin.

Around a decade before I was needed to come here from the village just opposite of the river. **I had more than 10 *bigha* of lands in that village but now I have lost almost all. Since I have lost most of my lands in river erosion, my life and livelihood is affected a lot.** At present, I have only 40 decimals land which is not eroded yet. This land, we used for crop cultivation. Since the production increased a lot than before with the modern system, our production increased a lot than before. I worked in this field. Not our required food crops partly comes from this field and partly comes from the market. After losing almost all lands, my elder two sons became workless in the village. To support the family, around 12 years before, both of them migrated to Dhaka. They are working in the garments. They support us financially. With their help, we can survive. From our childhood, we are struggling with climate change issue. We have witnessed severe floods in number of times. We lost not only our cultivable lands but also our residential land. We need to stay in governments' *khash* land which was primarily designed for a government hospital. Not only floods, we have experience of facing sudden storm, heavy rain, and extreme cold and even extreme hot. People of today are very concern. Before the floods, local people try to rescue the villagers. We never depend on government. We heard government help people financially but we never depend on them. We do hard work. My sons are my strength. They help to survive. In this village we have set up sanitary latrine and a tube-well also. We have never checked of arsenic level in water. We hope there is no problem of arsenic in our area. My wife takes care of the poultry and livestock. My sons help their mother in carrying grass and other feed for the livestock. With my sons' hard work, we are able to purchase refrigerator, television, fan and even mobile. If the nature supports us, we can bring our resilient life with our hard work, otherwise not.

Analysis of Case Study 1:

Although Md. Rabbil Ajij Talukdar lost all his cultivable land and homeland, his family was able to revert to a stable live by sending two sons for migration. This family primarily migrated with the full family from another village by losing all their properties in a river erosion. By migrating from their own village, they took shelter on government land which was supposed to be a land for a government hospital but as this land is just beside the terrible river Jamuna, the government cancelled their project of establishing a hospital in this region. Now many villagers like Ajij Talukdar live in government provided land untill it does not erode in the river.

Out of five children, Md Rabbil sent two elder sons to Dhaka city to work in garments and to support the family living in Meghai. With his sons' financial support, this family was able to buy a small piece of agro land, poultry and livestock. This family received partial support of food crops from this land. Additionally, support came from poultry and livestock which were taken care of by his wife. With the help of his migrated sons, this family was able to set up a latrine and tube well in their own place which was not possible by many families.

After losing all their resources in the river erosion, this family struggled a lot to get back a stable life. There were a number of government projects available to support the villagers. These projects were mostly political and worked to help the own followers and supporters. Md Rabbil was unfortunate as he had no political affiliation, thus he did not get any support from these projects.

Case Study 2: Ayesha Khatun, Meghai. Kajipur.Sirajganj.

I am Ayesha Khatun. I have no formal education. My original village is on the other side of the river. After losing our own residential land, we need to move in this place. We take shelter on the governments' place which is popularly known as hospitals land. I have two sons and a daughter.

Two of them have migrated to Dhaka for earning their livelihood. Only the younger son resides here with me. He is doing study in school. **My daughter is working in a garment of Dhaka, whereas my son is driving car for one of the garments in Dhaka.** My husband works as a labor in the rail station. **I am not passing time in leisure. I do support the family by working in the agricultural land of others and by taking care of the poultry and livestock in our own house**. These poultry and livestock support us in crisis time. Since, **it is not easy to manage food for the livestock; I need to spend hours to collect grass and other food for my livestock**. We have our own kitchen and room for our cattle. Two years before, my son and daughter migrated to Dhaka. **They help us financially in a regular basis**. We set up a latrine and a tube well to collect drinking water. We have planted few trees in my home yard. As we have electricity in our house, we purchase almost all the modern electronic devices including refrigerator, television and mobile. In our area we suffer mostly from river erosion. Though we are trying to get back our resilient but any time this new house can go inside the river. Our life is very uncertain with number of disasters. Every year we suffer from floods. Heavy winds, storm and extreme hot is also common in our area. We get back few lands from river as the char has risen but this land is not suitable for cultivation. This land has fallen as barren land. So we need to depend on market to purchase all our food products.

Analysis of Case Study 2

This family was found in a good economic condition. Two members from this family migrated to Dhaka and like many other migrants they work in the garments sector. As these migrants provided economic support regularly, this family was found with all modern accessories such as refrigerator, television and mobile.

This family was only a few of the blessed families, who can afford to continue electricity in that village. However, climate change and effects put this family in an acute situation like the others. Although they were capable of purchasing poultry and livestock but to support these, the natural source of foods were not available in their nearby areas. Interestingly, all the members of this family were found very hardworking. Even though their migrant members sent their parents the monthly expenses, yet all members work and try to earn within their capacity. Despite being hardworking, they struggled hard to maintain their family expenses as they cannot produce anything, rather they needed to depend on markets for eveything.

> **Case Study 3: Abdur Rashid (Khoka), Sariakandi, Bogra.**
>
> I did study up to 9 grade. Our family did not come here from any other place. We have our own land of residence. We have our own house in our own land. Beside our own residential land, we have our cultivable land also. We have around one and half *bigha* of lands for our agricultural production. Beside land, we have sewing machine which my wife use for tailoring purpose in a commercial way. Apart from these, we do have cattle and poultry as well. We have our own kitchen, toilet and tube-well also. We collect our drinking water from this tube-well which provides us arsenic free water.
>
> I can realize the effects of changed climate from my childhood. Almost every year we face number of natural hazards. However in last three years, we suffered mostly from river erosion and stormy rain. Conversely, this is also true that river erosion has reduced a lot than before as the government has taken proper steps. Since our land is very fertile, our production increased immensely than afore. After 1988's flood, I never witness a gigantic flood like that.

> I do work in my agricultural field. I have two sons. They migrated to Bogra town to work and support the family. One of them works as a driver in private bus and the other one works as a mechanic in a furniture shop. Both of them visit the village frequently and support the family as much as they can afford. So we can maintain our three times meal easily in a day. Government is supporting the vulnerable villagers. As we are not listed as the vulnerable villagers, we do not get that support. Actually we should not wait for that.

Analysis of Case Study 3:

This case was found as one of the luckiest farmers who had their own residential and cultivable lands. Even though the land could not produce enough crops required by this family, this land gave them mental comfort. Moreover, this piece of land made them socially dignified than those who did not have such lands. However, as this family could not afford enough to provide education for their sons, they had to work in the informal sector for earnings. For work, they sent their sons to the nearby town. The reasons for sending them to the nearby towns are that it was easy to communicate and the culture was familiar and known to them. This family faced poverty in their life but based on their social prestige and dignity, they did not need to be included in the government's list of vulnerable people.

> **Case Study 4: Ismail Hossain, Shaghata, Gaibanda**
>
> I did not get chance to do study. I have no formal education. I have no land, neither for my home nor for my livelihood. I had few *bigha* of cultivable lands which I lost in river bank erosion. This erosion does not occur all on a sudden. It took few years to erode. So having no alternative, I work as a labor in others agricultural field. This work is very seasonal work. Most of the time, I need to stay out of work.

> This area is very sandy during summer season. We cannot produce any crops except few pulses. In summer, we face great sufferings in drinking water as the water layer goes down. As I do not have own lands, my sons need to work to others land. In every year, they migrate to other villages for 3 to 6 months. They work there as the migrant labor in agriculture. Their earning help us to maintain our family in rest of the times of year which is really hard for our family. Very often I need to do loan from various sources like rich people and the bank. This is why, river erosion reduced a lot than before. But still this is the most devastating hazards for us. Beside this, extreme cold, rain and hot also create trouble for us. For financial aid, we do not wait for government. Government makes delay. They make short list based on network which I do not have. So we do not wait for their help.

Conclusion

Internal migration contributes a lot to the well-being of the families of origin which must be documented to sustain a country's development. Migration keeps a positive role in flowing resources to the families residing in the place of origin to minimize their threat of poverty. Migrants' behaviour towards their family of origin depends upon two issues, namely migrants' individual characteristics and characteristics of the migrants' household of origin. Type of family status, education, age, and type of work are the key factors that determine the behavioral pattern of the migrants. On the other hand, the number of family members, household income, and availability of resources such as amount of land owned, number of cattle and poultry, availability of tube well, toilet, and electricity determine the behavioral pattern of migrants.

CHAPTER 10

Discussion, Conclusion and Recommendations

1. Socio-demography and migration:

The demography of the villagers prompts them to make a decision on migration. Those who have their own lands, either for farming or residence, prefer not to move to the town whereas the landless, marginal poor have no other alternative than to move to other places for earning. Likewise, there is a relationship between gender and migration. It is found that the preferable member for migration is the father, son or the grandson. However, in some unusual cases such as absence of eligible male members or some progressive minded parents or if the family has acute poverty, the household allows the female members for migration. Age is found as another dominant factor for migration. For example, in all cases it is found that the young members of the family were sent for migration. Once they become old, in most cases, they prefer to come back to their own places. The literacy rate of these villagers along with their migrant members is very low. However, if any migrant with their hardship and visionary sight, obtains formal education, can manage better work and salary than others. For all cases, both the garments sector and construction sector were found as the most dependable sectors for work of migrant workers.

2. Migration throughout their life/ migration generation after generation:

In the research areas, the climate change and its effects are evident. Different types of disasters are seen throughout the year which are a result from climate change. The global temperature increased more than 2 degrees Celsius since 1974 (IPCC, 2007), which has serious impacts on glacier melting. This climate change leads to negative implications on the human society. Climate change with natural hazards such as floods, droughts, river erosions, and stormy winds are common in all three researched areas of the greater Northern areas. The long term negative effects of climate change shrink the alternatives of livelihoods in these areas. So for their earnings they move to other places depending on their capacities. Therefore, as they have very limited capacities they cannot move out of the country, instead they move to a nearer destination. Once they are able to manage some savings for their family maintenance, they come back to their place of origin and move again if they feel any crisis.

3. Chiefly internal movement seeking for a better standard of living

The inhabitants of these vulnerable areas are marginal people having no socio-economic capital. They do not have enough cultivable lands as most of them lost their lands in river erosions. These landless farmers do not have technical and certified knowledge of non-farming works. On the other hand, these people do not have the required assets to sell and conduct international migration. So, they prefer to go for internal and seasonal migration in the nearby urban, semi-urban and peripheral areas. However, all the inhabitants of the villages cannot migrate equally. Those who have few means such as strong social network or other means of communication, are able to move to other places. Similarly, few villagers were found who have assets like lands, cattle and homelands which are socially considered as the elite and respected people, and they usually do not prefer to move to other places. However, the elite send their next generation to town for higher studies and prefer them not to ever come back to the villages. This is because the elite are also in a threat to lose their properties in disasters and climate change. So, the trends of migration vary considerably depending on the vulnerabilities shaped by the individual, community and national level.

4. Typically with the expectation of eventually returning, again highlighting circular migration.

People of developing countries integrate the effects of climate change into their living strategies. Unavailability of assets and social capital restrict them to adapt and multiply their sources of earning. The prime intention of migration of the villagers was to earn their living. Therefore, after fulfilling their target of earning, they come back to their place of origin and migrate again whenever they are in lack of their living. Primarily, they migrate on a temporary basis and this process continues in a circulatory way. They prefer working as seasonal workers in the farming sector or other unskilled or semi-skilled working sectors. This type of migration is typically temporary, seasonal, internal and circulatory which proves that migration brings stability in an economic situation and reduces poverty in these vulnerable areas.

5. Voluntary or Distressed and forced migration:

Brown and Moore (1970) in their research talked about three stages of migration. In the first stage, the person faces problems to maintain his life in his place of residence and then he looks for alternatives. In the second stage of his migration, with that desire to migrate the person looks for the facilities available in the targeted place, and in the third stage the actual migration takes place. The nature of migration in these areas are voluntary and forced at a time but these stages of Brown and Moore were found true for these areas too. To adjust with the distressed, the villagers need to leave their home which creates a force to migrate. On the other hand, the villagers send their children to town for a better earning or living standard which can be considered as voluntary migration by the villagers.

Natural disasters and climate change forced the distressed villagers to migrate and these migrated villagers return to their own place once they are satisfied with their attained aid. Distressed migration results from local displacement as the immediate response to disasters and climate change. In most cases, it is found that villagers prefer to move to safer and nearer places in response to structural loss and destruction of efficacies to disasters. These people choose their destination based on their socio economic capital and with hope in mind that they may get back their previous place again once the immediate threat is overcome.

6. Environmental reasons more generally could possibly trigger a decision to move

Global climate change affects nature dependent inhabitants to look to migration forsurvival. To reduce the stress resulting from climate change and its effects, the human community adapts alternatives where migration is one of them. With climate change and its effects, some other socio-economic factors work together as the push and pull factors of the migration decision of these villagers. The effects of environmental change on their livelihood trigger their decision to move to other places. It is found that continuous and repetitive attack of natural disasters cause the gradual degradation of eco-system resulting in poor agricultural production and changing bio-diversity.

7. Migration and social network

Both the socialization and assimilation process from sociology is needed to conduct migration to internalize the new social positions. Social networks work as a mechanism for a successful assimilation in new social positions (Pescosolido, 1986). This is found true in this research. Those who have a strong social network can move to the new place and adjust themselves with ease.

As the villagers of these areas are poor economically, they fail to create enough social networks with others. Throughout their life, they struggle to look for their subsistence and to fight against the nature and climate to have a safer life. So, in general, all the villagers of the villages are weak in making social networks with others. However, among these villagers who have a minimum scale of social networking with others, they can look for migration and to change their future easily than those who have poor social networking. With a strong social network, some of them can manage work and shelter in town and can migrate comfortably than the other group of villagers.

8. The social valuation of migration and migrants in sending society is higher than in receiving society

The villagers who migrates from the village due to climate change and its effects receive social respect and importance in their own village. This is because they can send back money to their relatives which help them to solve many social and family problems.

This specific reason upgrades the social valuation of migrants in their own village. However, they do not get a similar social valuation in their new place. Whatever the type of migration, this migration cannot upgrade their social valuation in the receiving society. In most cases, the receiving society is unknown to them and does not bother about their personal identity. This new place only cares about the role that a migrant is supposed to perform. If they can perform their role properly, that new society values them.

9. Crisis of social identity in the place of receiving areas: psychological separation and competition in new areas.

Karl Marx's alienation theory is applicable in the life of migrants in receiving areas. The migrants may have a personal identity in their own village. They may be someone's son, someone's brother, someone's husband or someone's beloved one in their own society. But these identities have no social value in the receiving areas and at the beginning they will get a psychological shock in their new place. In the new place, no one cares about their old identity, here the only identity is what job and role they are performing. As the climate creates many changes in these villages along with other parts of the country, the tendency of the inhabitants to move to town and other suitable places for work and safety is higher than any other time. This trend creates huge pressure on the targeted areas of movement and forms a competition in every sector of the citizen's life. This situation puts an identity crisis in the new areas and a psychological separation from all kinds of their relations to survive in a competitive work environment.

Conclusions

Climate change and natural disasters already have a severe impact on the socio-economic lives of millions of people in Bangladesh. Changes to our climatic system, such as floods, droughts, and desertification and rainfall patterns impose huge stresses on the livelihood of the people of Bangladesh. Scientists warned that some parts of Bangladesh may become completely dilapidated due to environmental degradation and climate change. This situation places the inhabitants in a pressure socio-economically.

Most of the scientists and researchers stressed on the impacts of climate change and the increasing number of migration by the millions of people to towns and cities to create an urban life that is more suffocating and devastating. Researchers also triggered on the discrimination and inequality created from a climate induced migration. Unlike many others, throughout this book it is shown the necessity of migration, the push and pull factors of migration and the effects of migration in the lives of migrants and their families. To do this a framework is shown to show the relationship between various linking factors.

It is evident that climate change increases the number of human migration (Manou, et al; 2017). However, very few evidence is found where it is stated why, how, when and where do climatic vulnerable people move. These answers are given in this book.

There is always a debate in climate induced migration, whether this migration is voluntary or forced migration. This issue is also discussed in this book. In this book, the socio-political, demographic and economic issues are discussed to determine the factors that interplay a crucial role in climate induced migration.

Limitations

The limitations of this research-based book can be made a directory for future researchers. The following limitations are guidance for future research works.

1. The study covered only three villages from three different districts of Bangladesh. It could be a holistic research if more villages from more districts can be included.

2. This research includes a theoretical framework, conceptual framework and analytical tools from a sociological point of view. If other disciplines can use their own theories and tools for the same research objectives, the research can be an inclusive work.

Recommendations

Based on the obtained data, the following recommendations are made:

1. Government and non-government organizations should conduct more multidisciplinary researches before initiating and implementing policies for the people of this region.
2. Government should generate both agro and non- agro work opportunities for the people of this region. These people are skilled in agricultural works including cultivating, fishing and other farming. Scientific innovation is needed to grow crops in the changed climatic situation. Inhabitants need training and awareness programmes to be adapted with the changed situation and resilient themselves. Similarly, for non-agricultural works, they need training so that they can be skilled workforce in towns and cities.
3. Organizations which are working for the people of this region may introduce a "network hub" from where villagers may get the information on possible work opportunities in other towns and cities. Data shows social networking works as an important source to get work in towns and those who do not have strong personal and social networking, are not able to get work. In this situation, this "network hub" may help these people to get work.

REFERENCE

Arango, J. (2017). Theories of international migration. In International migration in the new millennium (pp. 25-45). Routledge.

Brown, L. A., & Moore, E. G. (1970). The intra-urban migration process: a perspective. Geografiska Annaler: Series B, Human Geography, 52(1), 1-13.

Cameron, M. P. (2018). Climate change, internal migration, and the future spatial distribution of population: a case study of New Zealand. Population and Environment, 39(3), 239-260.

Haug, S. (2008). Migration networks and migration decision-making. Journal of Ethnic and Migration Studies, 34(4), 585-605.

Hagen-Zanker, J. (2008). Why do people migrate? A review of the theoretical literature.

Intergovernmental Panel on Climate Change [IPCC]. (2014). Summary for policymakers. In C. B. Field, V. R. Barros, D. J. Dokken, K. J. Mach, M. D. Mastrandrea, T. E. Bilir, M. Chatterjee, K. L. Ebi, Y. O. Estrada, R. C. Genova, B. Girma, E. S. Kissel, A. N. Levy, S. MacCracken, P. R. Mastrandrea, and L. L.White (Eds.), Climate Change 2014: Impacts, Adaptation, and Vulnerability. Part A: Global and Sectoral Aspects. Contribution of Working Group II to the Fifth Assessment Report of the Intergovernmental Panel on

Climate Change. Cambridge University Press, Cambridge, United Kingdom and New York, NY, USA.

Klaiber, H. A. (2014). Migration and household adaptation to climate: a review of empirical research. Energy Economics, 46, 539-547.

United Nations Climate Change (2018). Article / 03 Jul, 2018. Climate Change is Driving Debt for Developing Countries.

Ravenstein, E. G. (1889). The laws of migration. Journal of the royal statistical society, 52(2), 241-305.

Massey, D. S., Arango, J., Hugo, G., Kouaouci, A., Pellegrino, A., & Taylor, J. E. (1993). Theories of international migration: A review and appraisal. Population and development review, 19(3), 431-466.

Tabor, A. S., & Milfont, T. L. (2011). Migration change model: Exploring the process of migration on a psychological level. International Journal of Intercultural Relations, 35(6), 818-832.

Sjaastad, L. A. (1962). The costs and returns of human migration. Journal of political Economy, 70(5, Part 2), 80-93.

Castro, L. J., & Rogers, A. (1984). Model migration schedules. A Simplified.

Uhlenberg, P. (1973). Noneconomic determinants of nonmigration: sociological considerations for migration theory. Rural Sociology, 38(3), 296.

Castles, S. (2003). Towards a sociology of forced migration and social transformation. Sociology, 37(1), 13-34.

Eakin, H. (2005). Institutional change, climate risk, and rural vulnerability: Cases from Central Mexico. World Development, 33(11), 1923-1938.

Sen, A. (1981). Ingredients of famine analysis: availability and entitlements. The quarterly journal of economics, 96(3), 433-464.

McLeman, R. A., & Hunter, L. M. (2010). Migration in the context of vulnerability and adaptation to climate change: insights from analogues. Wiley Interdisciplinary Reviews: Climate Change, 1(3), 450-461.

Van der Land, V., & Hummel, D. (2013). Vulnerability and the role of education in environmentally induced migration in Mali and Senegal. Ecology and Society, 18(4).

Braun, B., & Aßheuer, T. (2011). Floods in megacity environments: vulnerability and coping strategies of slum dwellers in Dhaka/Bangladesh. Natural hazards, 58(2), 771-787.

Burkart, K., Gruebner, O., Khan, M. M. H., & Staffeld, R. (2008). Megacity Dhaka-informal settlements, urban environment and public health. Geographische Rundschau, 4(1), 4-10.

Jabeen, H., Johnson, C., & Allen, A. (2010). Built-in resilience: learning from grassroots coping strategies for climate variability. Environment and Urbanization, 22(2), 415-431.

Mengistu, D. K. (2011). Farmers' perception and knowledge on climate change and their coping strategies to the related hazards: case study from Adiha, central Tigray, Ethiopia. Agricultural Sciences, 2(02), 138.

Alam, K., Shamsuddoha, M., Tanner, T., Sultana, M., Huq, M. J., & Kabir, S. S. (2011). The political economy of climate resilient development planning in Bangladesh. IDS Bulletin, 42(3), 52-61.

Ayers, J. M., & Huq, S. (2009). The value of linking mitigation and adaptation: a case study of Bangladesh. Environmental Management, 43(5), 753-764.

Huq, S., & Rabbani, G. (2011). Climate change and Bangladesh: policy and institutional development to reduce vulnerability. Journal of Bangladesh Studies, 13(1), 1-10.

Adger, W. N. (2006). Vulnerability. Global environmental change, 16(3), 268-281.

Barrios, S., Bertinelli, L., &Strobl, E. (2006). Climatic change and rural–urban migration: The case of sub-Saharan Africa. *Journal of Urban Economics, 60*(3), 357-371.

Brueckner, J., &Lall, S. (2015). Cities in Developing Countries: Fueled by Rural-Urban Migration, Lacking in Tenure Security, and Short of Affordable Housing. *Handbook of Regional and Urban Economics, 5.*

Cattaneo, C., &Massetti, E. (2015). Migration and Climate Change in Rural Africa.

Dessel, G. V. (2013). How to Determine Population and Survey Sample Size. https://www.checkmarket.com/2013/02/how-to-estimate-your-population-and-survey-sample-size/

Hunter, L. M., Luna, J. K., & Norton, R. M. (2015). Environmental Dimensions of Migration. *Annual Review of Sociology*, (0).

Iqbal, K., & Roy, P. K. (2015). Climate Change, Agriculture And Migration: Evidence From Bangladesh. *Climate Change Economics, 6*(02), 1550006.

Koenig, M. A., Ahmed, S., Hossain, M. B., &Mozumder, A. K. A. (2003). Women's status and domestic violence in rural Bangladesh: individual-and community-level effects. *Demography, 40*(2), 269-288.

Krejcie, R. V., & Morgan, D. W. (1970). Determining sample size for research activities.

Educational and Psychological Measurement, 30, 607-610.

Manou, D., Baldwin, A., Cubie, D., Mihr, A., & Thorp, T. (Eds.). (2017). *Climate Change, Migration and Human Rights: Law and Policy Perspectives.* Taylor & Francis.

Paul, S.K., Hossain, M. N. & Ray. S.K. (2013). Monga' in northern region of Bangladesh: a study on people's survival strategies and coping capacities. Rajshahi University journal of life & earth and agricultural sciences ISSN 2309-0960 Vol. 41: 41-56, 2013

Perch-Nielsen, S. L., Bättig, M. B., &Imboden, D. (2008). Exploring the link between climate change and migration. *Climatic Change, 91*(3-4), 375-393.

Plowman, A. (2015). Could the Effects of Climate Change be Profitable? A case study of climate induced migration into the Bangladeshi readymade garments industry.

Rehdanz, K., Welsch, H., Narita, D., & Okubo, T. (2015). Well-being Effects of a Major Natural Disaster: The Case of Fukushima. *Journal of Economic Behavior & Organization.*

Reuveny, R. (2007). Climate change-induced migration and violent conflict.*Political Geography, 26*(6), 656-673.

Saroar, M., &Routray, J. K. (2010). Adaptation in situ or retreat? A multivariate approach to explore the factors that guide the peoples' preference against the impacts of sea level rise in coastal Bangladesh. *Local Environment, 15*(7), 663-686.

Schipper, L., &Pelling, M. (2006). Disaster risk, climate change and international development: scope for, and challenges to, integration. *Disasters, 30* (1), 19-38.

Tibesigwa, B., &Visser, M. (2015). Small-scale Subsistence Farming, Food Security, Climate Change and Adaptation in South Africa: Male-Female Headed Households and Urban-Rural Nexus.

Uddin, A. M. K., (2009). Climate Change and Bangladesh. Seminar on Impact of Climate Change in Bangladesh and Results from Recent Studies. Organized by Institute of Water Modelling.

Unite For Sight. http://www.uniteforsight.org/global-health-university/importance-of-quality-sample-size

World Bank (2013). Planning, Connecting, and Financing Cities Now: Priorities for City Leaders. World Bank, Washington, DC.

Zug, S. (2006). Monga - Seasonal Food Insecurity in Bangladesh - Bringing the Information Together. *The Journal of Social Studies*, No. 111, July-Sept. 2006, Centre for Social Studies, Dhaka.

Weber, E. U. (2010). What shapes perceptions of climate change?. Wiley Interdisciplinary Reviews: Climate Change, 1(3), 332-342.

Mertz, O., Mbow, C., Reenberg, A., & Diouf, A. (2009). Farmers' perceptions of climate change and agricultural adaptation strategies in rural Sahel. Environmental management, 43(5), 804-816.

Capstick, S., Whitmarsh, L., Poortinga, W., Pidgeon, N., & Upham, P. (2015). International trends in public perceptions of climate change over the past quarter century. Wiley Interdisciplinary Reviews: Climate Change, 6(1), 35-61.

Weber, E. U. (2016). What shapes perceptions of climate change? New research since 2010. Wiley Interdisciplinary Reviews: Climate Change, 7(1), 125-134.

MacDonald, J. S., & MacDonald, L. D. (1964). Chain migration ethnic neighborhood formation and social networks. The Milbank Memorial Fund Quarterly, 42(1), 82-97.

Poros, M. V. (2001). The role of migrant networks in linking local labour markets: the case of Asian Indian migration to New York and London. Global networks, 1(3), 243-260.

Black, R., Adger, W. N., Arnell, N. W., Dercon, S., Geddes, A., & Thomas, D. (2011). The effect of environmental change on human migration. Global environmental change, 21, S3-S11.

Kron, W. (2013). Coasts: the high-risk areas of the world. Natural hazards, 66(3), 1363-1382.

Lucas, R. E. (1997). Internal migration in developing countries. Handbook of population and family economics, 1, 721-798.

McLeman, R., & Smit, B. (2006). Migration as an adaptation to climate change. Climatic change, 76(1-2), 31-53.

Black, R., Bennett, S. R., Thomas, S. M., & Beddington, J. R. (2011). Climate change: Migration as adaptation. Nature, 478(7370), 447.

Laczko, F., & Aghazarm, C. (2009). Migration, Environment and Climate Change: assessing the evidence. International Organization for Migration (IOM).

Mustari, S., & Karim, A. Z. (2017). Migration, an Alternative to Bring Resilience for Coastal Bangladesh: Napitkhali Village Experience. European Journal of Social Sciences, 54(3), 396-405.

Myers, G. M., & Papageorgiou, Y. Y. (1997). Efficient Nash equilibria in a federal economy with migration costs. Regional Science and Urban Economics, 27(4-5), 345-371.

Stern, N. (2006). The Stern Review on the Economic Effects of Climate Change (Report to the British Government). Cambridge University Press, Cambridge.

Henry, S., Schoumaker, B., & Beauchemin, C. (2004). The impact of rainfall on the first out-migration: A multi-level event-history analysis in Burkina Faso. Population and environment, 25(5), 423-460.

Hallegatte, S., Bangalore, M., Bonzanigo, L., Fay, M., Kane, T., Narloch, U., ... & Vogt-Schilb, A. (2015). Shock waves: managing the impacts of climate change on poverty. The World Bank.

Skoufias, E. (2012). The poverty and welfare impacts of climate change: quantifying the effects, identifying the adaptation strategies. The World Bank.

Faist, T., & Schade, J. (2013). Disentangling migration and climate change. Springer.

Meze-Hausken, E. (2000). Migration caused by climate change: how vulnerable are people inn dryland areas?. Mitigation and Adaptation Strategies for Global Change, 5(4), 379-406

Baker, J. L. (Ed.). (2012). Climate change, disaster risk, and the urban poor: cities building resilience for a changing world. The World Bank.

Pescosolido, B. A. (1986). Migration, medical care preferences and the lay referral system: A network theory of role assimilation. American Sociological Review, 523-540.

Parkins, N. C. (2010). Push and pull factors of migration. American Review of Political Economy, 8(2), 6.

Coccia, M. (2018). World-System Theory: A sociopolitical approach to explain World economic development in a capitalistic economy. Journal of Economics and Political Economy, 5(4), 459-465.

Todisco, E., Brandi, M. C., & Tattolo, G. (2003). Skilled migration: a theoretical framework and the case of foreign researchers in Italy.

Karim, M. R., Ampony, D., Muhammad, N., Noman, M. R. F., & Rahman, M. S. (2017). Factors affecting adopting of local adaptation options to climate change vulnerability. Journal of Science and Technology, 15, 1-8.

Coirolo, C., & Rahman, A. (2014). Power and differential climate change vulnerability among extremely poor people in Northwest Bangladesh: lessons for mainstreaming. Climate and Development, 6(4), 336-344.

Bhuyan, M. D. I., Islam, M. M., & Bhuiyan, M. E. K. (2018). A trend analysis of temperature and rainfall to predict climate change for northwestern region of Bangladesh. American Journal of Climate Change, 7(2), 115-134.

Shahid, S. (2011). Trends in extreme rainfall events of Bangladesh. Theoretical and applied climatology, 104(3-4), 489-499.

www.ingramcontent.com/pod-product-compliance
Lightning Source LLC
Chambersburg PA
CBHW030856180526
45163CB00004B/1600